THE BRITISH SCHOOL
OF BRUSSELS A.S.B.L.
Steenweg op Leuven 15b
1980 Tervuren

Multiple-Choice Questions in 'O' Level Chemistry

53

53

MULTIPLE-CHOICE QUESTIONS IN 'O' LEVEL CHEMISTRY

ROBIN A. H. HILLMAN, B.Sc., Dip. Ed.

*Head of Chemistry Department,
All Saints School, Bloxham*

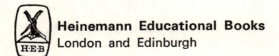

Heinemann Educational Books
London and Edinburgh

Heinemann Educational Books Ltd
LONDON EDINBURGH MELBOURNE AUCKLAND TORONTO
HONG KONG SINGAPORE KUALA LUMPUR NEW DELHI
IBADAN NAIROBI LUSAKA JOHANNESBURG
KINGSTON

ISBN 0 435 64321 5
Pupils ISBN 0 435 64320 7

© Robin A. H. Hillman 1971
First published 1971
Reprinted 1972, 1974, 1976

Published by Heinemann Educational Books Ltd
48 Charles Street, London W1X 8AH

Printed and bound in Great Britain by
Butler & Tanner Ltd, Frome and London

PREFACE

These questions are designed to help pupils studying the Nuffield 'O' Level Chemistry examination course. They are also suitable for students following the London and J.M.B. 'O' Level syllabuses and other modern courses. Each chapter might well be used as a test or set as homework on completing a particular topic. The questions would also be useful as revision for the 'O' Level examination. A little over one minute per question should be allowed.

A set of rubrics giving instructions for each section of questions is printed on a fold-out flap from the back of the cover. This flap should be opened out whenever the book is used.

As far as possible, SI units have been used and the recommendations of IUPAC on nomenclature have been followed. Certain deviations have, however, been made for convenience, e.g. some trivial names such as 'lime water' have been retained. The term 'atomic mass' has, however, been used throughout.

The questions have been collected over a period of three years and have been validated after use in tests and examinations at Bloxham School. It is hoped that most of the questions are original. The inspiration for some of the questions came from colleagues: in particular David Hood, Alan Griffin, Gordon Nelson, and the late Mark Atkinson. To all of these I am most grateful for allowing me to make use of their ideas. I should also like to thank the boys of Bloxham School who have participated in the testing of these questions and who have offered a number of constructive criticisms and suggestions. Gordon Nelson kindly read the manuscript and suggested several improvements, as did other colleagues. It is hoped that, as far as possible, all ambiguities and errors have been removed: the author would be glad to receive notification of any others which come to light.

Finally, I should like to thank my wife for typing the manuscript.

1971   R.A.H.H.

CONTENTS

	Page
PREFACE	v
TABLE OF APPROXIMATE ATOMIC MASSES	viii

1. Separation — 1
2. Heating Various Substances — 7
3. Air — 10
4. Burning — 15
5. Elements — 21
6. Competition Among the Elements — 24
7. Hydrogen — 27
8. Electricity and Chemistry — 31
9. Chemicals from Rocks — 34
10. Chemicals from the Sea — 37
11. Atoms — 41
12. An Important Gas — 48
13. The Periodic Table — 51
14. The Arrangement of Atoms in Elements — 58
15. Solids, Liquids, and Gases — 63
16. Electrolysis — 67
17. Reacting Quantities — 75
18. Rates of Reaction — 82
19. Equilibria — 87
20. Acids — 92
21. Molecules—Big and Small — 99
22. Ammonia — 103
23. Energy — 108
24. Radio-chemistry — 115

Table of Approximate Atomic Masses

Aluminium	27	Lead	207
Barium	137	Magnesium	24
Bromine	80	Manganese	55
Calcium	40	Mercury	200
Carbon	12	Nitrogen	14
Chlorine	35	Oxygen	16
Copper	64	Phosphorus	31
Helium	4	Potassium	39
Hydrogen	1	Silver	108
Iodine	127	Sodium	23
Iron	56	Sulphur	32

Volume of 1 Mole of Gas

Assume that under room conditions one mole of gas molecules occupies a volume of 24 dm^3.

Chapter 1

SEPARATION

* **Directions for Questions 1 to 15 (See Rubric A)**

Questions 1 to 8
    A. Chromatography
    B. Distillation
    C. Fractional distillation
    D. Evaporation
    E. Filtration

Which of the above processes would you use

1. to obtain the salt from salt water?

2. to obtain water from a solution of sugar in water?

3. to obtain xanthophyll from the green colouring matter in plants?

4. to remove solid impurities from water obtained from the sea?

5. to separate petrol from crude oil?

6. to separate sand from a mixture of sand and copper sulphate that has been shaken up with water?

7. to separate the different dyes from a sample of black ink?

8. to separate xylene (a liquid of boiling point 140 °C) and water from a mixture of the two liquids?

---

* **N.B.** The rubrics necessary for answering all the questions in this book will be found inside the flap of the back cover.

Questions 9 to 11

    A. Solute
    B. Solution
    C. Solvent
    D. Insoluble
    E. Saturated solution

Which of these terms describes

9. a liquid which extracts the red colouring matter from rose petals?

10. the liquid which remains when hot brine is cooled down until crystals of salt appear?

11. a red powder which, when stirred with alcohol and filtered, all remains on the filter paper?

Questions 12 to 15

From the list of pieces of apparatus:
A. A thermometer and a test-tube
B. An evaporating basin
C. A flask and a condenser
D. A separating funnel (a funnel fitted with a tap)
E. A crystallizing dish, a beaker, and a filter funnel

choose any *one* of A–E in each case which would best be used

12. to separate water and petrol

13. to determine whether or not some water contains a small amount of dissolved solids

14. to decide whether a liquid was pure alcohol or a mixture of alcohol and water

15. to purify crude copper sulphate

Directions for Questions 16 to 21 (See Rubric B)

16. A pupil was given a number of aqueous solutions and told to determine the pH of each using universal indicator paper. Here are his results:

Solution:	a	b	c	d	e	f	g
pH	3	4	7	5	8	9	6

    With which of the following mixtures might it be possible to make a solution with the same pH as c?
    A. a + b
    B. d + e
    C. e + f
    D. a + b + d
    E. b + d + g

Separation

17. Some dilute acid was put into a beaker and a few drops of litmus solution were added to it: the solution went red. When a solution of sodium hydroxide was added, the solution turned blue. Which of the following could be added to restore the red colour?
   A. water
   B. lime water
   C. litmus solution
   D. petrol
   E. vinegar

18. A scientist suspects that some canned peas contain a mixture of two green dyes, X and Y. He crushes the peas with water and makes a chromatogram with the liquid. He finds that the chromatogram consists of a single ring of dye X.
   He tries again using acetone as a solvent, and gets a single ring of dye Y. He then repeats the experiment, using alcohol as a solvent, and obtains a separation of the two dyes in his chromatogram.
   Which of the following statements is true?
   A. X is soluble in alcohol but Y is insoluble
   B. Y is soluble in water but X is insoluble
   C. Y is soluble in water but insoluble in acetone
   D. X is soluble in alcohol but insoluble in acetone
   E. X and Y are both soluble in acetone

19. A nurseryman's catalogue claims that the brilliant colour of a new rose he is offering for sale is due to the fact that its petals contain the same colouring matter as the flowers of the scarlet geranium. A chemist, anxious to test the truth of the claim, dissolves the colouring matter out of the petals of each flower in a suitable solvent. Which of the following methods would you advise him to use?
   A. To see if the solutions appear the same colour
   B. To find out if both solutions have the same pH
   C. To investigate whether both solutions act as indicators
   D. To see if the solutions can be distilled over at the same temperature
   E. To place a few drops of each solution on separate pieces of filter paper, add a few more drops of solvent, and watch how the colouring matter spreads out in each case.

20. When crude oil is distilled a number of fractions are collected. Which of these statements about the fractions is true?
   A. The first fraction has the lightest colour
   B. The first fraction has the highest boiling point
   C. The first fraction is the most viscous
   D. The first fraction burns least well
   E. The first fraction is a single pure substance

21. Rocksil wool, a material similar to asbestos wool, is placed in the bottom of a test-tube in which crude oil is to be heated. The reason for this is that
   A. it stops the oil from catching fire
   B. it must be added if it is desired to separate the oil into its fractions
   C. it absorbs the oil which can then be heated smoothly
   D. it prevents the test-tube from breaking when it is heated
   E. it prevents the oil from running out of the test-tube while the apparatus is being set up

Directions for Questions 22 to 24 (See Rubric C)

22. Good crystals of copper sulphate can be obtained by
    (i) heating copper sulphate until it melts, then allowing it to cool slowly
    (ii) allowing an aqueous solution of copper sulphate to evaporate slowly
    (iii) boiling off all the water from an aqueous solution
    (iv) boiling off the water from an aqueous solution until crystals just start to form, then allowing the remaining solution to cool

23. When an alum crystal is being grown by suspending it in a beaker containing a solution of the same substance
    (i) water is gradually evaporating from the solution
    (ii) more alum is being deposited on the crystal
    (iii) the solution must be saturated
    (iv) heating the solution will accelerate the growth of the crystal

24. The apparatus shown in Figure 1.1 is used to separate a mixture of alcohol and water. Which of the following statements is (are) true?

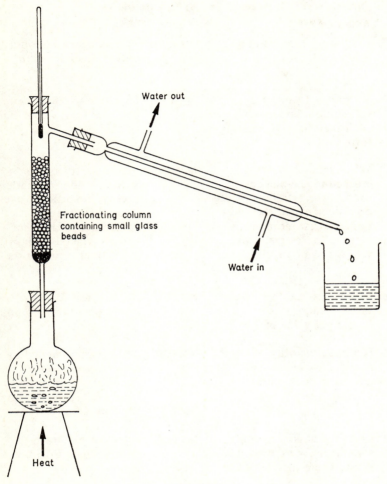

Figure 1.1

(i) The temperature in the fractionating column increases from top to bottom
(ii) The presence of glass beads helps the vapours to condense in the fractionating column.
(iii) The temperature of each fraction is measured by placing the bulb of the thermometer level with the side arm
(iv) The separation of the water and alcohol would be improved by using half the quantity of glass beads because the vapours would then reach the condenser more easily

Directions for Questions 25 to 30 (See Rubric D)

	Assertion		Reason
25.	A solution consists of a solute and a solvent	BECAUSE	in making a solution a solvent is dissolved in a solute
26.	When a hot saturated solution of copper sulphate is allowed to cool, crystals form in it	BECAUSE	copper sulphate is more soluble in hot water than in cold
27.	When spinach leaves are crushed in acetone and a drop of the extract is placed on a filter paper, the addition of a few more drops of acetone causes green and yellow rings to appear	BECAUSE	acetone extracts the various colouring matters present in spinach leaves and causes them to spread out to different extents when a drop of the extract is placed on filter paper
28.	A liquid is said to be pure water because it is found to have a pH value of 7	BECAUSE	the best test for the purity of water is to find its pH value
29.	If a large quantity of sugar is added to lemon juice, the acid taste disappears	BECAUSE	when sugar is added to lemon juice the sugar masks the taste of the lemon juice; the solution remains acidic
30.	Certain plants need acid soils; therefore they will thrive where the pH of the soil is 8.5	BECAUSE	values of pH greater than 7 indicate acidity

Chapter 2

HEATING VARIOUS SUBSTANCES

Directions for Questions 1 to 5 (See Rubric A)

Questions 1 to 5
  A. Cobalt chloride crystals
  B. Copper foil
  C. Iodine crystals
  D. Red lead oxide
  E. Zinc oxide

Which of these substances

1. produces water on heating?

2. turns yellow on heating and returns to white on cooling?

3. increases in mass on heating?

4. completely vaporizes on heating?

5. leaves a yellow residue on cooling after heating?

Directions for Questions 6 to 11 (See Rubric B)

6. Which of the following describes a pure substance?
   A. A substance which leaves a residue when filtered
   B. A liquid which has a pH of 7
   C. A compound containing mercury and oxygen only
   D. A liquid which starts to boil at 80 °C, the last trace of liquid boiling away at 100 °C, leaving no residue
   E. Tap water

7. 'The substance starts reddish-pink, and on heating it gives off a vapour which condenses on the sides of the test-tube. The substance turns blue. If, on cooling, water is added to the residue, it returns to its original colour.'
   To which of the following substances does the above description apply?
   A. Cobalt chloride crystals
   B. Copper sulphate crystals
   C. Iodine crystals
   D. Iron pyrites
   E. Zinc oxide

8. Which of the following would *not* change in mass when heated in air?
   A. Copper
   B. Magnesium
   C. Nichrome wire
   D. Red lead oxide
   E. Potassium permanganate

9. Which of the following would you expect to gain in mass when heated in air?
   A. Red lead oxide
   B. Glass
   C. Copper carbonate
   D. Magnesium
   E. Platinum

10. When a sample of sand was heated its mass was found to decrease The most likely reason for this is that
    A. the sand gave off a gas when heated
    B. the sand changed into another substance when heated
    C. the sand lost its water of crystallization when heated
    D. the sand was damp
    E. sand has a smaller mass when hot

11. A piece of copper was heated in a Bunsen flame. On cooling, it was noticed that the copper had become blackened. Which of the following best expresses the change that has taken place?
    A. The copper has taken something out of the flame
    B. The copper has become coated with a thin layer of soot
    C. The copper has combined with part of the air to form a new substance
    D. The copper has changed colour, but no new substance has been formed
    E. The copper has taken something out of the air, but no new substance has been formed

Directions for Questions 12 to 15 (See Rubric C)

12. Water of crystallization is
    (i) the water left behind when a substance crystallizes from solution
    (ii) the water absorbed by a crystal when it is formed
    (iii) the water used to make a saturated solution in which a crystal is grown
    (iv) the water given off when hydrated compounds are heated

13. Cobalt chloride paper is
    (i) an indicator used to distinguish between acid and alkali
    (ii) used to detect the presence of water in liquids
    (iii) used to show that a sample of water is pure
    (iv) kept in a desiccator

14. Copper sulphate is an important chemical substance. Which of the following are true statements of its properties?
    (i) If an iron nail is placed in a solution of copper sulphate, it becomes coated with a layer of copper
    (ii) Anhydrous copper sulphate is a blue crystalline substance
    (iii) In a solution of copper sulphate, the copper sulphate is present as the solute
    (iv) If copper sulphate is heated gently, a gas is produced which is not easily condensed

15. Which of the following tests is (are) suitable for investigating whether a liquid is pure water?
    (i) Test the pH of the liquid
    (ii) Test the liquid with anhydrous copper sulphate
    (iii) Test the liquid with cobalt chloride paper
    (iv) Test the boiling point of the liquid

Directions for Questions 16 to 20 (See Rubric D)

Assertion		**Reason**
16. Blue copper sulphate crystals lose water on heating	BECAUSE	anhydrous copper sulphate is a good test for the presence of water
17. Dilute sulphuric acid would be expected to turn anhydrous copper sulphate blue	BECAUSE	indicators can be used to detect the presence of acids
18. When a pure liquid has begun to boil, its temperature continues to rise while it is boiling	BECAUSE	the temperature at which a liquid boils depends on how strongly it is heated
19. To find out if there is any change in mass when magnesium ribbon and copper foil are heated in air, the metals are heated separately in open crucibles	BECAUSE	when magnesium and copper are heated in air, they form white and black compounds respectively, both of which are solids
20. When a piece of copper foil is folded over and the edges made flat by hammering, it is found that, on heating in a Bunsen flame, only the outside of the foil is blackened	BECAUSE	when an envelope of copper foil is heated in a Bunsen flame, the temperature inside the envelope does not rise sufficiently high to cause blackening of the inside of the foil

Chapter 3

AIR

Directions for Questions 1 to 3 (See Rubric A)

Questions 1 to 3

    A. 0.06   B. 0.6   C. 6   D. 15   E. 30

Choose from the above list, the letter indicating the correct answer to each of the following:

1. The density, in grammes per cubic decimeter, of a certain gas, 50 cm³ of which weigh 0.3 g

2. The decrease in volume of air when pyrogallol, a substance which absorbs only oxygen, is shaken with 30 cm³ of air

3. The volume of oxygen used up when, in a syringe experiment, 5.20 g of copper are converted into 5.24 g of copper oxide (assume density of oxygen = 1.33 g dm⁻³)

Directions for Questions 4 to 7 (See Rubric B)

4. Which of the following is the most accurate statement about the composition of air?
   A. It is an equal mixture of nitrogen and oxygen
   B. It is a mixture consisting of about 4/5 nitrogen and 1/5 oxygen
   C. It is a mixture consisting of about 4/5 nitrogen and 1/5 oxygen with a little water vapour and carbon dioxide
   D. It is a mixture consisting of almost 4/5 nitrogen, about 1/5 oxygen and small quantities of other gases
   E. It is a compound consisting of 4/5 nitrogen and 1/5 oxygen

5. Air that has had oxygen removed from it is sometimes described as being 'inert'. The meaning of the word 'inert' is best explained as
   A. chemically inactive
   B. does not burn
   C. puts out a glowing splint
   D. describing a gas which does not turn heated copper black
   E. insoluble

Air   11

6. A test-tube containing a few grammes of potassium permanganate was heated. Oxygen was given off and collected in a gas jar full of water (the gas jar having been previously placed in a position to collect the gas). Which of the following conclusions is *incorrect*?
   A. The gas collected was not pure oxygen as some air was also displaced from the test-tube
   B. The gas collected also contained water vapour
   C. The gas collected could not be very soluble
   D. The gas actually collected in this experiment would differ in volume from the gas produced by heating the potassium permanganate, for the former volume would also include the volume of air displaced from the test-tube. (Assume that all the volumes were measured under identical conditions)
   E. The potassium permanganate had undergone a chemical change

7. Priestley first obtained oxygen in 1774 by
   A. burning magnesium
   B. heating mercury oxide
   C. heating red lead oxide
   D. removing the copper from copper oxide
   E. removing the other gases from a sample of air

Directions for Questions 8 to 11 (See Rubric C)

8. The apparatus shown in Figure 3.1 was set up to investigate the heating of copper in air. Air was passed from one syringe to the other over the copper which was placed in two heaps in the silica

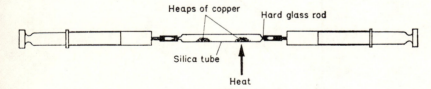

Figure 3.1

tube connecting the syringes. The syringes were both of 100 cm³ capacity and at the start of the experiment both were half-full of air. The tube was heated at the point shown in the diagram. It was found that
   (i) both heaps of copper became black
   (ii) the total amount of air left at the end was 80 cm³
   (iii) the copper was decomposed by the heat
   (iv) the gas left in the syringes after the experiment extinguished a lighted match

9. A boy decided to investigate whether any change in mass occurred when red lead oxide was heated in an open test-tube. Before carrying out the experiment, he listed the precautions that are necessary. His list should have included:
   (i) The red lead oxide must be well dried before the initial weighing
   (ii) The red lead oxide must be heated gently in order to avoid cracking the test-tube
   (iii) The test-tube must be allowed to cool before re-weighing the tube and its contents
   (iv) The reaction must be carried out in a fume cupboard because lead compounds are poisonous

10. A group of boys each heated some potassium permanganate in a test-tube and collected the gas evolved in a gas jar marked with graduations corresponding to every 10 cm^3 of gas produced.

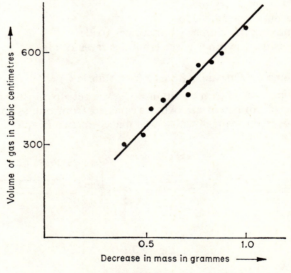

Figure 3.2

A graph was constructed (Figure 3.2) to show how the decrease in mass of the potassium permanganate that occurred on heating was related to the volume of gas evolved. On the graph were plotted the results of ten different boys as shown.
   (i) A graph of the type shown, for this and for any similar experiment, must always pass through the origin

(ii) A point on the straight line would give a more reliable picture of the relationship between the decrease in mass and the volume of gas produced than would a particular boy's results
(iii) If the graph were continued to give the decrease in mass corresponding to the production of 1000 cm³, the decrease in mass would be numerically equal to the density of the gas under the conditions of the experiment, measured in grammes per cubic decimetre
(iv) The ten boys could not have worked very carefully; otherwise all the points would have fallen on the straight line

11. Lavoisier heated mercury in the apparatus shown in Figure 3.3.

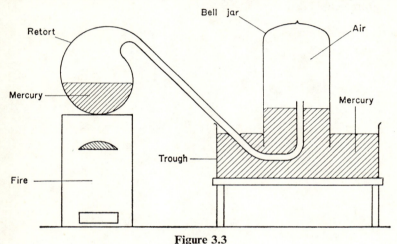

Figure 3.3

Amongst the results of his experiments were:
(i) The volume of air in the bell jar decreased during the heating
(ii) When a splint was placed in the bell jar after the experiment was over, it went out
(iii) Particles of a red substance formed on the surface of the mercury in the retort
(iv) Particles of a red substance formed on the surface of the mercury in the trough

Directions for Questions 12 to 15 (See Rubric D)

Assertion		Reason
12. When copper is heated in air the change is temporary	BECAUSE	when copper is heated in air, a compound of copper and oxygen is formed

13. When strongly heated, red lead oxide loses oxygen BECAUSE when strongly heated, red lead oxide changes colour

14. Oxygen is tested by re-lighting a glowing wood splint BECAUSE oxygen is the active part of air; so that substances would be expected to burn better in pure oxygen than in air

15. Mercury must always be heated in a well-ventilated fume cupboard BECAUSE mercury vapour is very poisonous

Chapter 4

BURNING

Directions for Questions 1 to 6 (See Rubric A)

Questions 1 to 3
    A. Carbon
    B. Copper
    C. Iron
    D. Magnesium
    E. Sulphur

From these substances which

1. burns in air to form a white powder?

2. burns with a blue flame?

3. burns in air to give a product that on mixing with water has a pH of 10–11

Questions 4 to 6
    A. Acidic gas
    B. Acidic solid
    C. Alkaline liquid
    D. Alkaline solid
    E. Neutral solid

Which of these terms best describes the nature of the oxide formed when the following are heated in oxygen?

4. sulphur

5. magnesium

6. carbon

16 Multiple-Choice Questions in 'O' Level Chemistry

Directions for Questions 7 to 11 (See Rubric B)

7. The apparatus shown in Figure 4.1 is set up to investigate the change in mass when a candle is burnt. Which of the following 'purposes' is *not* correct?

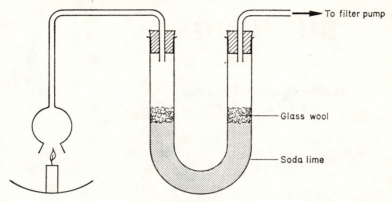

Figure 4.1

   A. The purpose of the filter pump is to draw the gaseous products through the apparatus
   B. The purpose of the soda lime is to absorb the gaseous products produced by the burning of the candle
   C. The purpose of the glass wool is to absorb any material not absorbed by the soda lime
   D. The purpose of the bent thistle funnel is to ensure that as much of the gaseous products as possible will pass through the apparatus
   E. The purpose of carrying out the experiment, once with a lit candle and once with an unlit candle, is to see how much of the increase in mass of the soda lime is due to moisture and carbon dioxide present in air

8. A certain solid element was heated in a sealed vessel containing air, so that nothing could get into or out of the vessel. The element burned to form a solid oxide. Which one of the following would you expect?
   A. The mass of gas in the vessel would not alter
   B. The mass of solid in the vessel would not alter
   C. The total mass of the vessel and its contents would be more after the experiment than it was before
   D. The total mass of the vessel and its contents would be the same after the experiment as it was before
   E. The total mass of the vessel and its contents would be less after the experiment than it was before

9. In Figure 4.2, when air is blown through the tube A, air bubbles out through B. When air is sucked in through the tube A the air is

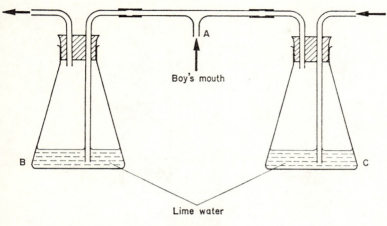

Figure 4.2

sucked in through C. B and C both contain lime water. A boy sucks and blows alternately through A. Which of these changes will occur?

A. The lime water is unaffected
B. The lime water in B turns milky but never in C
C. The lime water in C turns milky but never in B
D. The lime water turns milky in both but very much more quickly in B than in C
E. The lime water turns milky in both but very much more quickly in C than in B

10. When red lead is heated it changes colour and a gas is given off. This is an example of

A. burning
B. combustion
C. decomposition
D. physical change
E. refining

11. A piece of white phosphorus is burnt inside the apparatus shown in Figure 4.3. Some universal indicator solution is added to the water. Oxides of phosphorus are soluble in water, and phosphorus

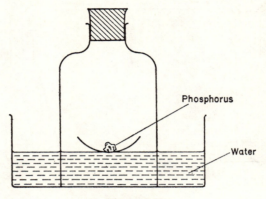

Figure 4.3

is a non-metal. Which of the following will be observed after a short time?
A. The solution shows an acidic reaction and the inside level has remained the same
B. The solution shows an acidic reaction and the inside level has risen
C. The solution shows an alkaline reaction and the inside level has risen
D. The solution shows a neutral reaction and the inside level has risen
E. The solution shows a neutral reaction and the inside level has remained the same

Directions for Questions 12 to 15 (See Rubric C)

12. When a jet of town-gas is burnt inside a flask, a mist appears on the inside of the flask.
    (i) If the moisture is tested with universal indicator it is found to be acid
    (ii) A piece of cobalt chloride paper turns from blue to pink when it is rubbed against the side of the flask
    (iii) The mist is produced because the oxygen in the air combines with the substances present in town gas
    (iv) The mist appears on the side of the flask because the flask is colder than the air above the flame

Burning

13. When a candle burns inside an upturned beaker
   (i) carbon dioxide is produced
   (ii) water is produced
   (iii) the candle goes out when all the oxygen in the beaker has been used up
   (iv) the candle goes out when the only constituent of the original air left in the beaker is nitrogen

14. An experiment is carried out in which a piece of charcoal is heated in air and then immediately put into a test-tube of oxygen. Correct notes about the experiment include which of the following?
   (i) The charcoal glowed more brightly in oxygen than in air
   (ii) When the charcoal burned a gas was formed which, when bubbled through lime water, turned the lime water milky
   (iii) When a few drops of universal indicator were added to the test-tube after the charcoal had burned and the test-tube was shaken thoroughly, the colour of the indicator showed a pH of 6
   (iv) After the burning the gas in the test-tube was also tested with a piece of moist red litmus paper. This was found to turn blue.

15. Under which of the conditions in Figure 4.4 will iron rust?

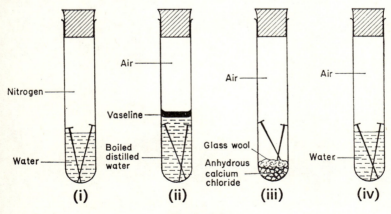

Figure 4.4

Directions for Questions 16 to 20 (See Rubric D)

	Assertion		Reason
16.	To test gas for carbon dioxide it is bubbled through lime water	BECAUSE	lime water goes cloudy on exposure to air for some time
17.	The gas from burning sulphur does not turn lime water milky	BECAUSE	sulphur combines with oxygen when it is burnt in air
18.	1 gramme of magnesium would gain more in mass when completely burnt in an atmosphere of pure oxygen than when completely burnt in air	BECAUSE	only one-fifth (approximately) of the air is oxygen
19.	When magnesium is burnt in oxygen the product is an alkaline oxide	BECAUSE	oxygen is an alkaline gas
20.	Iron will not rust in dry air	BECAUSE	oxygen is needed for rust to form

Chapter 5

ELEMENTS

Directions for Questions 1 to 3 (See Rubric A)

Questions 1 to 3

Which of these elements:

A. aluminium
B. carbon
C. gold
D. sodium
E. sulphur

1. is a yellow solid which does not conduct electricity?

2. is a non-metallic element which conducts electricity in one of its forms?

3. is a metal which melts easily, and burns with a yellow flame?

Directions for Questions 4 to 9 (See Rubric B)
4. Which of the following is the best definition of an element?
   A. A pure substance
   B. A substance made by decomposing a compound
   C. A substance which burns in air forming an oxide
   D. A substance which cannot be split into anything simpler by chemical means
   E. A substance which cannot be split into anything smaller by chemical means

5. One of the general properties of metals is their ductile nature. This means that
   A. they are hard
   B. they are strong
   C. they have a high density
   D. they can be rolled into thin sheets
   E. they can be drawn out into wires

6. Which of the following facts is *not* true of the element sodium?
   A. It is soft and easily cut with a knife
   B. It is stored in liquid paraffin
   C. It burns easily
   D. Its density is lower than that of water
   E. It is a poor conductor of heat and electricity

7. A piece of calcium was burnt in oxygen. Which of the following statements would be correct?
   A. It burned with a yellow flame and the oxide dissolved in water turning red litmus blue
   B. It burned with a blue flame and the oxide produced was acidic
   C. It burned with a bright greenish-blue flame forming a white woolly substance sometimes referred to as 'Philosopher's Wool'
   D. It burned with a bright-red flame and the oxide produced a solution in water which turned universal indicator purple
   E. It burned with a dazzling white flame and dense white fumes of a choking gas were produced

8. When burning a piece of calcium in air it is usual to wrap the piece in asbestos paper before picking it up in tongs. The reason for this is that
   A. the reaction is extremely vigorous if the calcium is not wrapped in this way
   B. the asbestos paper prevents a reaction between the tongs and the calcium
   C. the asbestos paper prevents small bits of calcium from breaking off and falling on to the bunsen burner
   D. the asbestos paper absorbs any molten calcium
   E. the use of asbestos paper prevents heat being conducted away by the tongs, making it possible to heat the calcium at a higher temperature

9. The oxides of some elements give alkaline pH values when mixed with water and tested with universal indicator. Which of the following oxides would be the most alkaline?
   A. Aluminium oxide
   B. Iron(III) oxide
   C. Lead(II) oxide
   D. Magnesium oxide
   E. Zinc oxide

Elements

Directions for Questions 10 to 12 (See Rubric C)

10. Substances which are elements
    (i) cannot be decomposed into two or more different substances by heating
    (ii) include earth, air, fire, and water
    (iii) can react only by combining with other substances
    (iv) do not occur naturally: they have to be extracted from ores

11. Compounds have the following properties:
    (i) They are made from more than one element
    (ii) Their composition is fixed
    (iii) Compounds are produced as a result of a chemical change taking place
    (iv) Their properties are similar to those of the elements from which they are made

12. When a substance X was heated strongly it gave off a gas Y which was collected in a gas jar. A burning splint put into this gas instantly went out. From these observations it can be concluded that
    (i) X was an element
    (ii) the gas Y was oxygen
    (iii) the residue was acidic
    (iv) X lost weight when it was heated

Directions for Questions 13 to 15 (See Rubric D)

Assertion		Reason
13. Elements which conduct electricity are not decomposed by it	BECAUSE	elements are single 'pure' substances which cannot be split up into anything simpler
14. When a piece of lead foil is placed on a crucible lid and heated a yellow coloration is observed	BECAUSE	lead foil on heating strongly decomposes to give a yellow compound
15. When sulphur is heated in a gas jar of oxygen and the product shaken with universal indicator, the indicator shows that the substance formed is acidic	BECAUSE	non-metallic elements such as sulphur form acidic oxides when they are heated in oxygen

Chapter 6

COMPETITION AMONG THE ELEMENTS

Directions for Questions 1 to 4 (See Rubric A)

Questions 1 to 4
- A. Iron filings
- B. Iron(III) oxide
- C. Magnesium
- D. Magnesium oxide
- E. Zinc

Which of these substances would

1. react most vigorously with copper(II) oxide?

2. react with dry aluminium powder in the thermit reaction?

3. be the strongest reducing agent?

4. be the most likely to be reduced by carbon?

Directions for Questions 5 to 8 (See Rubric B)

5. It was found that zinc powder was much more reactive than a piece of zinc foil when both were heated in air. This is because
   A. the state of division of a substance greatly affects its reactivity
   B. zinc powder is chemically different from zinc foil
   C. zinc powder is much less pure than zinc foil; it is the presence of these impurities which gives the zinc enhanced reactivity
   D. the layer of oxide on the surface of the foil could not have been removed. If the surface of the foil had been rubbed with sandpaper, the foil would then have been as reactive as the powder
   E. the zinc powder was heated on a piece of asbestos paper which helped to concentrate the heat on the powder

Competition Among the Elements

6. A student made the following observations in the laboratory:
   (a) Clean copper metal did not react with lead nitrate solution
   (b) Clean lead metal dissolved in silver nitrate solution and crystals of silver appeared
   (c) Clean silver metal did not react with copper(II) nitrate solution

   The order of decreasing strength as reducing agents of the three metals is shown to be:
   A. Copper, lead, silver
   B. Copper, silver, lead
   C. Lead, silver, copper
   D. Lead, copper, silver
   E. Silver, lead, copper

7. Zinc is a more reactive metal than iron, which is more reactive than tin, which, in turn, is more reactive than lead.
   Which of the following statements is therefore *not* true?
   A. Iron can displace lead from lead(II) oxide
   B. Lead can displace iron from iron(III) oxide
   C. Tin can displace lead from lead(II) oxide
   D. Zinc can displace tin from tin(IV) oxide
   E. Zinc can displace iron from iron(III) oxide

8. If the oxide of an element X is reduced by an element Y and the oxide of Y is reduced by an element Z, then
   A. any element which reduces the oxide of X will reduce the oxides of Y and Z
   B. any element which reduces the oxide of Y will be reduced by Z
   C. the oxide of Z will be reduced by X
   D. X can never reduce the oxide of an element that is reduced by Z
   E. Y cannot reduce the oxide of an element which cannot be reduced by Z

Directions for Questions 9 to 11 (See Rubric C)

9. Magnesium is placed above copper in the reactivity series because
   (i) magnesium burns fiercely in air whereas copper does not
   (ii) magnesium decomposes carbon dioxide but copper does not
   (iii) magnesium reacts with copper oxide but copper does not react with magnesium oxide
   (iv) magnesium is produced when copper is placed in a solution of magnesium sulphate

10. The metal tin is placed below iron in the reactivity series but is above lead. Tin can be
    (i) produced from its ore, cassiterite, an oxide of tin, by heating with carbon
    (ii) added to a solution of silver nitrate to displace the silver since a piece of silver foil placed in lead nitrate solution does not react
    (iii) used under the same conditions as lead in a reaction with chlorine, when it is seen to react more vigorously
    (iv) converted to its oxide by burning in air; the oxide on shaking with water turns universal indicator paper red

11. Carbon is more commonly used to reduce metal oxides than magnesium is. The reason for this is that
    (i) carbon does not catch fire as readily as magnesium
    (ii) carbon is a non-metal and magnesium is a metal
    (iii) carbon is higher in the reactivity series than magnesium
    (iv) magnesium is much more expensive than carbon

Directions for Questions 12 to 15 (See Rubric D)

Assertion		Reason
12. When the thermit reaction is carried out using a mixture of chromium(III) oxide and aluminium, the aluminium is said to reduce the chromium(III) oxide to chromium	BECAUSE	reduction means reducing the mass of a substance
13. Magnesium will reduce zinc oxide	BECAUSE	magnesium has a stronger affinity for oxygen than zinc has
14. A mixture of iron powder and magnesium oxide burns with a brilliant white flame when heated	BECAUSE	when iron powder is heated with magnesium oxide, the iron removes oxygen from the magnesium oxide
15. Lead can be placed above carbon in the reactivity series	BECAUSE	carbon will react with lead(II) oxide to form lead and carbon dioxide

Chapter 7

HYDROGEN

Directions for Questions 1 to 4 (See Rubric A)

Questions 1 to 4
- A. Rocksil wool (similar to asbestos wool) test-tube clamped horizontally fitted with bung and short glass tube
- B. Bent thistle funnel, filter pump and test-tube with side arm
- C. Conical flask with side arm, thistle funnel and delivery tube
- D. Gas supply, U-tube and test-tube with small hole near the bottom end
- E. Two gas syringes and connecting tube

Which of the above sets of apparatus would be most useful for

1. preparing hydrogen from a metal and dilute acid?

2. converting lead(II) oxide into lead using coal gas?

3. investigating the reaction between magnesium and steam?

4. investigating the product formed when hydrogen is burnt in air?

Directions for Questions 5 to 9 (See Rubric B)

5. Hydrogen can be produced by passing steam over iron filings in the apparatus shown in Figure 7.1. A number of precautions must be taken. Which of these is *not* necessary?

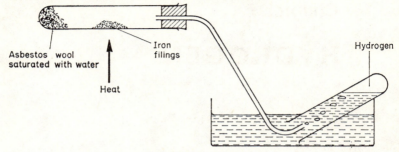

Figure 7.1

A. After the heating is stopped the delivery tube must be removed from under the surface of the water
B. Asbestos wool must be treated with respect, avoiding the breathing in of any small particles of the substance
C. The first one or two test-tubes of gas should be discarded if it is desired to collect only pure hydrogen
D. Because hydrogen is less dense than air, the delivery tube and the test-tubes used for collecting the gas must be at a higher level than the test-tube containing the iron filings, so that the gas can rise into the collecting tube
E. The asbestos should not be heated directly. The steam is generated by moving the bunsen backwards and forwards between the iron and the asbestos wool

6. Zinc reacts with dilute acid displacing hydrogen but copper does not. This indicates that
A. zinc is less reactive than copper
B. copper is found above zinc in the reactivity series
C. zinc is higher in the reactivity series than hydrogen
D. copper is inert
E. zinc is higher in the reactivity series than hydrogen but copper is lower

7. Steam reacts with hot aluminium. Hydrogen reacts with copper(II) oxide. Steam does not react with hot copper.
   On this evidence, between which one of the following pairs of substances is there likely to be a reaction?
A. Copper and aluminium oxide
B. Aluminium and copper(II) oxide
C. Copper(II) oxide and steam
D. Aluminium oxide and steam
E. Aluminium oxide and hydrogen

Questions 8 and 9

8. The equipment shown in Figure 7.2 is assembled in order to reduce copper(II) oxide with hydrogen gas. Hydrogen gas produced in generator A is passed over hot copper oxide in the combustion tube B.

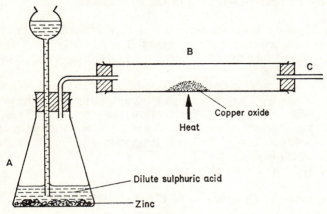

Figure 7.2

Tiny droplets of a clear liquid collect on the inside of tubes B and C. To test whether the liquid is water, or at least contains water, the best of the following methods would be to
A. add anhydrous copper sulphate and look for a colour change
B. smell the liquid
C. taste the liquid
D. add some litmus solution and look for a colour change
E. add universal indicator

9. Assume that the liquid identified in the last question is found to be water. Possibly water has been carried into the tube B with the hydrogen from the generator A. In order to test the validity of this explanation it would be best to
A. heat tube B further
B. heat generator A
C. add a calcium chloride drying tube to the right of C
D. add a calcium chloride drying tube to the left of B
E. try to produce hydrogen by the reaction of zinc with a different acid

Directions for Questions 10 and 11 (See Rubric C)

10. The following are some of the properties of hydrogen:
    (i) It is insoluble in water
    (ii) It supports combustion
    (iii) It will reduce lead(II) oxide to lead
    (iv) It forms about 80% of the atmosphere

11. The chemical name for water is hydrogen oxide. Evidence that this is its chemical nature includes the following fact(s):
    (i) Magnesium reacts with steam to form magnesium oxide and hydrogen
    (ii) A mist appears on the inside of a dry test-tube of hydrogen mixed with air when a lighted splint is applied to the mixture
    (iii) When natural gas (which may be regarded as 100% methane, a compound of carbon and hydrogen only) is burnt and the products cooled to room temperature, a liquid is obtained which boils at 100 °C
    (iv) Water is not an element

Directions for Questions 12 to 15 (See Rubric D)

Assertion		Reason
12. When steam is passed over heated magnesium contained in a glass tube, black markings are seen on the walls of the tube	BECAUSE	magnesium reacts with steam to form a dark-coloured substance, magnesium oxide
13. When hydrogen is prepared in the laboratory by the reaction of zinc with dilute sulphuric acid, a little copper sulphate solution is added to speed up the reaction	BECAUSE	hydrogen is produced by the reaction between dilute sulphuric acid and copper sulphate solution
14. When a jet of hydrogen is burned in air, an explosion soon results	BECAUSE	hydrogen and air mixtures are liable to be explosive
15. The major industrial use of hydrogen is in filling balloons	BECAUSE	hydrogen is the least dense of all gases

Chapter 8

ELECTRICITY AND CHEMISTRY

Directions for Questions 1 to 4 (See Rubric A)

Questions 1 to 4
- A. Distilled water
- B. Potassium iodide
- C. Sodium
- D. Sugar
- E. Sulphur

Choose from these substances the one which

1. conducts electricity when solid and when molten

2. does not conduct electricity when solid but does conduct electricity when molten

3. does not conduct electricity when solid, nor when in aqueous solution

4. gives a different product at the cathode when electrolysed in a molten state to that produced when it is electrolysed in aqueous solution

Directions for Questions 5 to 8 (See Rubric B)

5. An electrolyte can best be described as a substance which
   - A. conducts electricity
   - B. generates electricity
   - C. allows electricity to pass through it
   - D. conducts electricity when molten
   - E. conducts electricity and is decomposed by it

6. When a solution of copper sulphate is electrolysed, the product at the cathode is
   - A. copper
   - B. oxygen
   - C. hydrogen
   - D. sulphate
   - E. sulphuric acid

7. The electrolysis of a certain liquid resulted in the formation of hydrogen at the cathode and chlorine at the anode. Which one of the following could it be?
   A. A solution of copper chloride in water
   B. A solution of sodium chloride in water
   C. A solution of sulphuric acid in water
   D. Pure water
   E. None of the substances A–D

8. Two carbon rods are connected to a 6 volt battery and a bulb is also included in series in the circuit. The rods are inserted in some sodium chloride kept molten by a fierce Bunsen flame. The bulb lights up. The Bunsen flame is then slightly lowered and the bulb is seen to glow less brightly than before. The reason for this is that
   A. a non-conducting gas is produced by the electrolysis of the molten salt
   B. the carbon rods are no longer hot enough to conduct electricity well
   C. once the sodium chloride has melted, the stronger the heating the more brightly the bulb will glow
   D. all of the sodium chloride has solidified
   E. part of the sodium chloride has solidified

Directions for Questions 9 to 11 (See Rubric C)

9. Electrolysis is the splitting up of a substance by means of an electric current.
   Which of the following statements about electrolysis are correct?
   (i) Electrolysis is one of the most important methods of obtaining a metal from its ore
   (ii) In electrolysis the electrode connected to the positive of the electricity supply is termed the cathode
   (iii) In order for electrolysis to take place, it is necessary that the substance being electrolysed should be in liquid form, either molten or in solution
   (iv) When a solution of potassium iodide is electrolysed, potassium and iodine are produced

10. Industrially, copper is purified by electrolysis. Which of the following statements is (are) likely to apply to the process?
    (i) A thin strip of pure copper is used as the cathode
    (ii) A block of impure copper is used as the anode
    (iii) The electrolyte is copper sulphate solution
    (iv) An a.c. supply is used

Electricity and Chemistry

11. A piece of copper foil and a piece of magnesium ribbon are dipped into some sulphuric acid placed in a beaker. It is found that
    (i) if the pieces of metal are connected by a wire, a current flows between the magnesium and the copper
    (ii) the magnesium ribbon gradually dissolves in the acid
    (iii) chemical energy is being changed into electrical energy
    (iv) replacing the copper by a metal higher up in the reactivity series, for example iron or zinc, will provide a larger voltage

Directions for Questions 12 to 15 (See Rubric D)

Assertion		Reason
12. Mercury is an example of an electrolyte	BECAUSE	any liquid which conducts electricity is called an electrolyte
13. Sulphuric acid diluted with distilled water is a poor conductor of electricity	BECAUSE	pure water is a bad conductor of electricity
14. When copper chloride solution is electrolysed, copper is produced on the cathode; but when potassium chloride solution is electrolysed, no potassium appears to be produced at the cathode	BECAUSE	a solution of copper chloride conducts electricity; but a solution of potassium chloride does not
15. When an object is to be chromium-plated, it is used as the cathode in a suitable electrolysis cell	BECAUSE	metals are deposited on the cathode during electrolysis

Chapter 9

CHEMICALS FROM ROCKS

Directions for Questions 1 to 3 (See Rubric A)

Questions 1 to 3
- A. Calcium carbonate
- B. Calcium oxide
- C. Copper carbonate
- D. Iron(II) oxide
- E. Iron(II) sulphide

Which of these substances is also known as

1. chalk?

2. malachite?

3. quicklime?

Directions for Questions 4 to 9 (See Rubric B)

4. When malachite is heated, a black powder is formed. A student believes that this substance could be either carbon or copper(II) oxide. Which of the following tests would *not* help him to distinguish between the two substances?
   - A. Attempting to burn the powder in a gas jar of oxygen
   - B. Dissolving the powder in dilute sulphuric acid
   - C. Dissolving the powder in water
   - D. Mixing with magnesium powder and heating
   - E. Passing hydrogen gas over the heated powder

5. A substance is being reduced in a stream of coal gas. It is contained in a hard glass test-tube with a small hole at one end, the issuing gas being burnt at the hole. Which of the following instructions for this experiment is correct?
   - A. Clamp the test-tube near its centre
   - B. Turn on the gas supply and immediately apply a light to the hole
   - C. Adjust the gas supply so that the flame of the burning gas is not less than 5 cm high
   - D. There is no need to heat the substance: sufficient heat will be produced by the burning of the gas at the hole
   - E. When the reaction is complete, stop heating the substance but continue to pass the gas through the tube until it is cold

6. Which of the following is not chemically identical with all the others?
   A. Marble
   B. Chalk (from rocks)
   C. Limestone
   D. Quicklime
   E. Calcium carbonate

7. An insoluble powdered white substance was added to a little dilute hydrochloric acid. A gas was evolved which turned lime water milky. This white substance was also heated in a roaring Bunsen flame for a few minutes. The product was still a white solid but, when added to water, heat was given out whereas the initial substance was unaffected by water. The original powdered white substance could have been
   A. copper(II) carbonate
   B. lead(II) carbonate
   C. limestone
   D. magnesium oxide
   E. quicklime

8. An important calcium compound is slaked lime. Its properties include which of the following?
   A. It can be made by heating calcium in dry air
   B. It is unaffected by heat
   C. It is insoluble in water
   D. Its chemical name is calcium hydroxide
   E. When hydrochloric acid is added to a little of the substance and a drop of the resulting mixture placed in a Bunsen flame, the flame turns yellow

9. A student is investigating a substance which is known to be a carbonate. His results are listed below. Which of his experiments would you advise him to repeat?
   A. Analysis of the compound showed that it contained a metal, carbon and oxygen
   B. On adding dilute acid to a little of the substance, it effervesced vigorously
   C. On heating gently, the substance gave off a gas which relit a glowing splint
   D. On heating strongly, the substance gave off a gas which turned lime water milky
   E. The substance was soluble in water: it could not therefore be copper carbonate

Directions for Questions 10 to 12 (See Rubric C)

10. When iron(III) oxide is heated with carbon,
    (i) the iron(III) oxide is reduced
    (ii) the product is magnetic
    (iii) the product evolves hydrogen on treatment with dilute sulphuric acid
    (iv) steel is formed

11. Limestone is a cheap, readily available material. It is
    (i) used in a finely ground state for removing the acidity from soils
    (ii) mixed with water to produce a solution called lime water which is used for detecting carbon dioxide
    (iii) used in the blast furnace
    (iv) unaffected by dilute acids

12. Quicklime is a compound of calcium and oxygen. Which of the following statements about quicklime is (are) true?
    (i) It produces an acid solution when mixed with water
    (ii) It is produced when calcium is burnt in air
    (iii) If a wire is held in a Bunsen flame after dipping in a mixture of quicklime and concentrated hydrochloric acid, the flame turns a reddish colour
    (iv) It is produced by gently warming limestone

Directions for Questions 13 to 15 (See Rubric D)

Assertion		Reason
13. Brass is stronger than pure copper; it is therefore known as an alloy	BECAUSE	an alloy is a mixture of two or more metals; an alloy is usually stronger than the metals from which it is composed
14. Lime water is a test for carbon dioxide	BECAUSE	carbon dioxide is given off on heating limestone strongly
15. When a piece of marble is heated strongly its mass decreases	BECAUSE	marble loses oxygen when heated

Chapter 10

CHEMICALS FROM THE SEA

Directions for Questions 1 to 6 (See Rubric A)

Questions 1 to 6
- A. Barium chloride
- B. Hydrochloric acid
- C. Potassium chloride
- D. Silver nitrate
- E. Sulphuric acid (concentrated)

From the list of substances A–E, choose the one which

1. is used for testing a solution to find out if it contains a carbonate

2. is used for testing a solution to find out if it contains a chloride

3. is used for testing sea water to find out if it contains sulphates

4. gives a lilac-coloured flame

5. is obtained most easily when sea water is evaporated

6. on reacting with manganese(IV) oxide and the residue left by evaporating sea water to dryness, produces a reddish-brown vapour

Directions for Questions 7 to 12 (See Rubric B)

7. Which of the following compounds (all of which are present in sea water), would be the first to appear as a solid if a volume of sea water was evaporated?
   - A. Calcium carbonate
   - B. Calcium sulphate
   - C. Magnesium sulphate
   - D. Potassium chloride
   - E. Sodium chloride

8. A boy is investigating a substance whose name he does not know. He finds that when the substance is treated with dilute hydrochloric acid it effervesces, forming a gas which turns lime water milky. If a piece of nichrome wire is dipped into the acidic solution and then held in a Bunsen flame, the flame does not change colour. He concludes therefore that the substance is
   A. magnesium oxide
   B. potassium chloride
   C. potassium carbonate
   D. zinc chloride
   E. zinc carbonate

9. Which of the following statements about chlorine is *not* true?
   A. It is a poisonous yellow-green gas
   B. It is used for treating swimming bath water
   C. It is a more reactive element than iodine, being able to displace iodine from potassium iodide solution
   D. it burns in air or oxygen
   E. In industry, it is produced by the electrolysis of brine

10. Identify the use of bromine which is stated incorrectly:
    A. In the manufacture of 1,2-dibromoethane which is added to petrol to remove lead formed in the cylinders by decomposition of the 'anti-knock' present in petrol
    B. In the manufacture of flame retardants
    C. In the manufacture of medicines, especially sedatives
    D. In the manufacture of compounds used in making photographic film
    E. In the manufacture of compounds added to table salt to ensure that sufficient of the element is present in the body, since a deficiency leads to thyroid trouble

11. At one point in the process used for obtaining iodine from seaweed, the iodine present is extracted by shaking with a suitable solvent. The main purpose of this is to
    A. separate the iodine from impurities which are insoluble in the solvent used
    B. show that iodine is present in the mixture
    C. obtain a solution of iodine; iodine crystals are dangerous and therefore it is normally kept in solution
    D. stop the iodine from vaporizing
    E. cause a chemical change to take place

12. The solvent used in extracting iodine from seaweed is
    A. dilute sulphuric acid
    B. hydrogen peroxide
    C. starch solution
    D. tetrachloromethane
    E. water

Directions for Questions 13 to 16 (See Rubric C)

13. In performing a flame test
    (i) the substance being investigated is first moistened with hydrochloric acid
    (ii) sodium salts are recognized by an intense yellow flame
    (iii) calcium salts produce a brick red colouring
    (iv) potassium salts give a flame which is yellow when viewed through blue glass

14. A chemist is analysing a white powder. He finds that it gives a colour to a Bunsen flame after moistening with hydrochloric acid. In solution it gives a blue colour with Magneson reagent, a white precipitate with dilute nitric acid and silver nitrate solution, and a white precipitate with barium nitrate solution. On this evidence *alone*, which of the following mixtures might this substance be?
    (i) Magnesium sulphate and potassium chloride
    (ii) Magnesium chloride and sodium carbonate
    (iii) Magnesium chloride and sodium sulphate
    (iv) Potassium sulphate and sodium carbonate

15. When sea water is to be investigated, a convenient method of doing so is to evaporate the water in stages and to examine the products produced at each stage. Which of the following are advantages of this method?
    (i) It allows a partial separation of the substances present in sea water as a result of their differing solubilities
    (ii) It provides a means of concentrating the substances in sea water so that they may be more readily identified
    (iii) It permits a study of the differing crystal shapes of the substances present in sea water
    (iv) It prevents decomposition of the substances which would take place if the evaporation were carried out continuously

16. The electrolysis of sea water of different concentrations yields a variety of products. Correct statements about the electrolysis of sea water include:
    (i) When sea water (not concentrated) is electrolysed, chlorine is given off at the anode
    (ii) When sea water (not concentrated) is electrolysed, hydrogen is given off at the cathode
    (iii) When a concentrated solution of sea water is electrolysed, bromine is produced at the anode
    (iv) When a concentrated solution of sea water is electrolysed, sodium is produced at the cathode

Directions for Questions 17 to 20 (See Rubric D)

	Assertion		Reason
17.	Sodium chloride crystals are cubic in shape	BECAUSE	sodium chloride crystals are different in shape to those of magnesium chloride
18.	Starch-iodide paper is used as a test for chlorine	BECAUSE	chlorine reacts with starch to form a blue substance
19.	Fluorine is not a member of the halogen family of elements	BECAUSE	fluorine is so extremely reactive a gas that it cannot be used in school chemistry laboratories
20.	When gaseous bromine is bubbled through a solution of potassium iodide, the solution turns brown	BECAUSE	gaseous bromine is brown in colour

Chapter 11

ATOMS

Directions for Questions 1 to 6 (See Rubric A)

Questions 1 to 6

Choose from the following

A. 0.25 D. 2
B. 0.5   E. 4
C. 1

the figure representing the answer to each of the following questions:

1. The mass in grammes of 0.25 moles of oxygen atoms, O

2. The number of moles of carbon atoms, C, present in 3 g of carbon

3. The number of moles of hydrogen atoms, H, present in 9 g of water

4. The number of moles of calcium atoms, Ca, having the same number of atoms as 0.5 moles of silver atoms, Ag

5. The number of grammes of helium having the same number of atoms as 0.5 moles of silver atoms, Ag

6. The number of moles of mercury atoms, Hg, which combine with 1 mole of chlorine atoms, Cl, experiment having shown that 7.0 g of chlorine combine with 40 g of mercury

Directions for Questions 7 to 18 (See Rubric B)

7. Carbon dioxide gas is about 1½ times as dense as nitrogen gas. Suppose that a gas jar of nitrogen was placed on top of a gas jar of carbon dioxide with the open ends together. Which of the following would you expect to have happened after a while?
   A. The two gases would not have mixed at all
   B. Some of the carbon dioxide would have moved into the gas jar containing nitrogen, but none of the nitrogen would have moved into the gas jar containing carbon dioxide
   C. Some of the nitrogen would have moved into the gas jar containing carbon dioxide but none of the carbon dioxide would have moved into the gas jar containing nitrogen
   D. Some of each gas would have moved into the gas jar containing the other
   E. The two gases would have reacted with each other

8. Which one of the following statements is true?
   A. Equal masses of elements contain equal numbers of atoms
   B. A mole of atoms is the number of atoms present in 1 g of any element
   C. One mole of atoms of any element contains the same number of atoms
   D. The mass in grammes of one mole of any element is always a whole number
   E. It is necessary to use a mole as the standard unit of amount of an element because it is not possible to find the real masses of atoms

9. You have probably carried out an experiment to determine the thickness of an oil film on water.
   Which one of the following is the most reasonable conclusion that can be drawn from the oil drop experiment?
   A. Oil is composed of a vast number of separate atoms
   B. Oil is composed of very small particles, the size of which is given by the thickness of the film
   C. The film is so thin that oil must be one continuous substance
   D. If oil is composed of small particles, their maximum size in one direction is the thickness of the film
   E. If oil is composed of small particles, their size in one direction cannot be less than the thickness of the film

10. There are 10^9 nanometres in one metre and therefore 10^7 nanometres in one centimetre. One drop of a pure oil has a volume of 0.05 cm³ and the oil does not evaporate readily. The particles in the oil have a diameter of 2 nanometres.

   The drop of oil will form a film with an area of
   A. 25 cm²
   B. 250 cm²
   C. 2500 cm²
   D. 250 000 cm²
   E. 2 500 000 cm²

11. John Dalton first introduced the idea of atomic mass in the early nineteenth century. He regarded the atomic mass of an element as
   A. the number of times heavier an atom of that element is than hydrogen is
   B. the number of times heavier that element is than hydrogen is
   C. the mass of an atom of that element compared with the mass of an atom of hydrogen
   D. the number of times heavier an atom of that element is compared with a particle of hydrogen
   E. the mass of an atom of that element

12. Which of the following has the same number of atoms as 0.2 g of hydrogen?
   A. 0.2 g of oxygen
   B. 0.2 mole of chlorine atoms, Cl
   C. 0.5 mole of nitrogen atoms, N
   D. 8 g of bromine
   E. 8 g of chlorine

13. Which of the following masses of sodium has three times as many atoms as there are in two moles of potassium?
   A. 46 g   B. 69 g   C. 78 g   D. 117 g   E. 138 g

14. How many moles of silver atoms are displaced by one mole of magnesium atoms if 1.20 g of magnesium displace 10.8 g of silver from a solution of silver nitrate?
   A. 0.05   B. 0.5   C. 1   D. 2   E. 9

15. 0.75 g of a metal X was found to react with 0.32 g of oxygen. The atomic mass of X is 75. The formula of the compound formed was
   A. X_2O
   B. XO
   C. XO_2
   D. X_2O_3
   E. X_3O_2

16. When some oxide of mercury was decomposed into mercury and oxygen, it was found that 25 g of mercury and 1 g of oxygen were produced. The formula for the oxide is therefore
    A. HgO
    B. HgO_2
    C. Hg_2O
    D. Hg_2O_3
    E. Hg_3O_2

17. The formula for lead iodide is written as $PbI_2(s)$. Which of the following is *not* correct?
    A. The state symbol (s) implies that the lead iodide is a solid
    B. The formula of the substance must be derived from experiment
    C. The formula represents one gramme mole of lead iodide
    D. The formula represents 334 g of lead iodide
    E. 0.05 mole of lead atoms, Pb, would combine with 0.10 mole of iodine atoms, I

18. Copper(II) oxide may be reduced using the method shown in Figure 11.1. When all the copper oxide has been reduced to copper

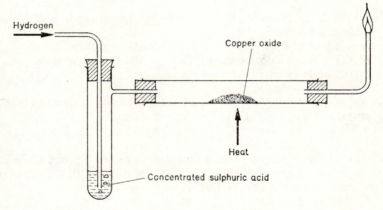

Figure 11.1

the combustion tube is cooled. During the cooling, a slow stream of hydrogen continues to be passed through the tube. The most important reason for this is that the passage of hydrogen
A. helps to cool the tube
B. prevents the tube from cooling too quickly
C. ensures that any water formed during the reaction is driven out of the tube
D. prevents oxidation of the warm copper
E. lessens the risk of an explosion by ensuring that air does not enter the tube while it is still hot

Directions to Questions 19 to 23 (See Rubric C)

19. In the oil-drop experiment, a substance such as stearic acid is dissolved in a suitable solvent and a drop of the solution is placed on the surface of water covered with a fine powder. It is necessary
  (i) to use a solvent which evaporates very quickly
  (ii) to use a solvent which does not itself form a film
  (iii) to assume that the oil film is one particle thick
  (iv) to use a very dilute solution of the substance

20. The atomic mass of an element expressed in grammes tells us:
  (i) the mass in grammes of one atom of that element
  (ii) the mass of the Avogadro constant of atoms of the element
  (iii) the mass of the element which will react with oxygen
  (iv) the mass of one mole of atoms of that element

21. 96 g of magnesium contain twice as many atoms as
  (i) 2 moles of calcium atoms, Ca
  (ii) 0.25 g of lead
  (iii) 32 g of oxygen
  (iv) 48 g of carbon

22. It was found that when 13.70 g of red lead oxide was heated, 0.32 g of oxygen were produced. The residue on reduction gave 12.42 g of lead. It can be concluded from this that
  (i) 6 moles of lead atoms, Pb, are contained in 1370 g of red lead oxide
  (ii) 128 g of oxygen are contained in 1370 g of red lead oxide
  (iii) the simplest formula for the residue is PbO
  (iv) when red lead oxide is heated it loses one quarter of its total amount of oxygen

23. The experiment illustrated in Figure 11.2 was designed as a method for determining the formula of water; the water being produced by

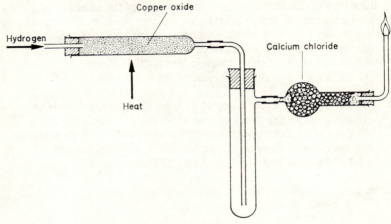

Figure 11.2

passing hydrogen over copper(II) oxide. Which of the following amendments to the diagram would you recommend?
  (i) Bubbling the hydrogen through concentrated sulphuric acid before allowing it to enter the combustion tube
  (ii) Placing some concentrated sulphuric acid in the side-arm test-tube
  (iii) Placing tufts of glass wool near each end of the combustion tube to hold the copper oxide in place
  (iv) Immersing the side-arm test-tube in an ice–water mixture

Directions for Questions 24 to 30 (See Rubric D)

Assertion		Reason
24. Dalton suggested that chemical combination took place between small whole numbers of atoms	BECAUSE	Dalton suggested that the smallest part of an element which could exist independently was an atom
25. When a bottle of ether is opened in a room, the smell of ether vapour gradually reaches all parts of the room	BECAUSE	when a bottle of ether is opened, draughts carry the vapour round the room; in a draught-free room the smell of ether would not be detected away from the vicinity of the bottle

Assertion		Reason
26. If a mixture of hydrogen and air is left for some time, it will gradually separate out with hydrogen as the top layer	BECAUSE	hydrogen is less dense than air
27. When ammonia gas and hydrogen chloride vapour are released at opposite ends of a long glass tube, a white ring of ammonium chloride appears nearer to the ammonia than to the hydrogen chloride	BECAUSE	ammonia gas diffuses more rapidly than hydrogen chloride
28. The volume of a mole of atoms of any element is the same	BECAUSE	a mole of atoms of an element always contains the same number of atoms
29. The symbol assigned to an element is used to represent one mole of atoms of that element	BECAUSE	the formula given to a compound depends on the relative numbers of moles of atoms of the elements combining together to form that compound
30. The formula of magnesium oxide is written MgO	BECAUSE	it has been found by experiment that 24 g of magnesium combines with 16 g of oxygen to form magnesium oxide

Chapter 12

AN IMPORTANT GAS

Directions for Questions 1 to 3 (See Rubric A)

Questions 1 to 3
    A. Copper(II) oxide   D. Potassium permanganate
    B. Hydrogen chloride   E. Sodium chloride
    C. Iron(II) chloride

Choose the appropriate letter for each of the substances described below:

1. This substance in solution forms chlorine gas when reacted with an oxidizing agent
2. This substance is the main constituent of rock salt
3. This substance is an oxidizing agent commonly used in the preparation of chlorine

Directions for Questions 4 to 8 (See Rubric B)

4. The chemical name for 'salt gas' is
   A. hydrochloric acid   C. chlorine   E. HCl
   B. hydrogen chloride   D. hydrogen oxide

5. When 'salt gas' is passed over iron, it reacts with it and another gas is produced which
   A. causes a glowing splint to re-ignite
   B. bleaches litmus paper
   C. burns
   D. turns lime water cloudy
   E. extinguishes a lighted splint

6. When 'salt gas' is passed over heated iron, in addition to a gas being formed
   A. the iron melts
   B. the iron vaporizes
   C. the iron catches fire and forms a red compound
   D. the iron reacts to give glistening crystals
   E. the iron rusts

An Important Gas

7. In an experiment to convert iron into iron(II) chloride by heating the metal in 'salt gas', it was found that 2.8 g of iron yielded 6.3 g of iron(II) chloride.
    The formula for iron(II) chloride must therefore be;
    A. Fe_3Cl
    B. Fe_2Cl
    C. $FeCl$
    D. $FeCl_2$
    E. $FeCl_3$

8. Which of the following is *not* a property of chlorine?
    A. It fumes in moist air
    B. It gives an acidic response to universal indicator
    C. It is yellowish-green in colour
    D. It extinguishes a glowing splint
    E. It does not burn

Directions for Questions 9 to 11 (See Rubric C)

9. 'Salt gas' is a name sometimes used for the gas which
    (i) turns lime water milky
    (ii) bleaches moist litmus paper
    (iii) forms when common salt is heated
    (iv) is produced by burning hydrogen in chlorine

10. Hydrogen and chlorine may be combined by passing the gases backwards and forwards between two syringes over a heated platinum spiral. It is necessary that
    (i) the apparatus be completely dry
    (ii) the apparatus be cooled before the final volume of gas is read
    (iii) the apparatus be protected from direct sunlight
    (iv) twice as much hydrogen as chlorine be used

11. The density of hydrogen is 0.09 g dm^{-3} and of chlorine 3.20 g dm^{-3}. An experiment showed that x cm^3 of hydrogen reacted with x cm^3 of chlorine to give $2x$ cm^3 of salt gas. It therefore follows that 2 g of hydrogen either react with or produce

    (i) $\dfrac{1000 \times 2}{0.09}$ cm^3 of chlorine

    (ii) $\dfrac{2000 \times 2}{0.09}$ cm^3 of 'salt gas'

    (iii) $\dfrac{3.20 \times 2}{0.09}$ g of chlorine

    (iv) $\dfrac{3.20}{0.09} \times 2 \times 35$ moles of chlorine atoms, Cl

Directions for Questions 12 to 15 (See Rubric D)

	Assertion		**Reason**
12.	Potassium permanganate reacts with a concentrated solution of 'salt gas' in water to give hydrogen	BECAUSE	'salt gas' contains hydrogen
13.	If it is desired to prepare a sample of hydrogen chloride gas from rock salt and concentrated sulphuric acid, the gas may be collected by displacement of water	BECAUSE	gases are often collected by displacement of water so as to obtain a sample of gas free from air
14.	When hydrogen chloride is made by passing hydrogen and chlorine over heated platinum, if the apparatus is somewhat damp, the volume of gas produced is slightly greater than expected	BECAUSE	hydrogen chloride fumes in moist air
15.	The masses of gases which react together are in a simple ratio	BECAUSE	the volumes of gases which react together are in a simple ratio

Chapter 13

THE PERIODIC TABLE

Directions for Questions 1 to 9 (See Rubric A)

Questions 1 to 5

Consider the following sets of elements, identified by their atomic numbers:
A. 2, 10, 18, 36, 54, 86
B. 3, 11, 19, 37, 55, 87
C. 4, 12, 20, 38, 56, 88
D. 5, 6, 7, 8, 9, 10
E. 9, 17, 35, 53, 85

1. Which set of elements is not a group of the periodic table?

2. Which set is called the alkali metal group?

3. Which set is called the halogen group?

4. In which set of elements is there at least one which is solid, one which is liquid and one which is a gas at room temperature?

5. Which set of elements has the highest atomic volumes?

Questions 6 to 9

The following are five common metals;
A. Copper
B. Iron
C. Lead
D. Magnesium
E. Sodium

Choose from the above five metals the metal described in each of the following statements:

6. The melting point of the metal is lower than the boiling point of water

7. The metal, when heated, reacts with iodine vapour to form a reddish-brown substance

8. The metal dissolves in dilute sulphuric acid giving off hydrogen and forming a coloured solution

9. This metal in powdered form is used to obtain silicon from sand

THE PERIODIC TABLE OF THE ELEMENTS

I	II											III	IV	V	VI	VII	O
						1 H											2 He
3 Li	4 Be											5 B	6 C	7 N	8 O	9 F	10 Ne
11 Na	12 Mg											13 Al	14 Si	15 P	16 S	17 Cl	18 Ar
19 K	20 Ca	21 Sc	22 Ti	23 V	24 Cr	25 Mn	26 Fe	27 Co	28 Ni	29 Cu	30 Zn	31 Ga	32 Ge	33 As	34 Se	35 Br	36 Kr
37 Rb	38 Sr	39 Y	40 Zr	41 Nb	42 Mo	43 Tc	44 Ru	45 Rh	46 Pd	47 Ag	48 Cd	49 In	50 Sn	51 Sb	52 Te	53 I	54 Xe
55 Cs	56 Ba	57 La	72 Hf	73 Ta	74 W	75 Re	76 Os	77 Ir	78 Pt	79 Au	80 Hg	81 Tl	82 Pb	83 Bi	84 Po	85 At	86 Rn
87 Fr	88 Ra	89 Ac	104 Ku	105													

58 Ce	59 Pr	60 Nd	61 Pm	62 Sm	63 Eu	64 Gd	65 Tb	66 Dy	67 Ho	68 Er	69 Tm	70 Yb	71 Lu
90 Th	91 Pa	92 U	93 Np	94 Pu	95 Am	96 Cm	97 Bk	98 Cf	99 Es	100 Fm	101 Md	102 No	103 Lr

Directions for Questions 10 to 19 (See Rubric B)

10. If the elements sodium, magnesium, aluminium, silicon, phosphorus, sulphur, chlorine, and argon are arranged in that order, the series is called
    A. a family
    B. a group
    C. a period
    D. a transition series
    E. the reactivity series

11. In the family of elements which includes lithium, sodium, and potassium,
    A. the element with the greatest atomic mass is the least reactive
    B. the atomic mass of potassium is between the atomic masses of lithium and sodium
    C. the density of the element is always greater than the density of water
    D. all react with water to form the gas oxygen
    E. all burn in air with a coloured flame

12. Rubidium is an element in the same group of the periodic table as lithium and sodium. It is likely to be a metal which is
    A. hard, with a high melting point, and which reacts slowly with water
    B. soft, with a high melting point, and very reactive with water
    C. soft, with a low melting point, and very reactive with water
    D. hard, with a low melting point and very reactive with water
    E. soft, with a low melting point, and which reacts slowly with water

13. When it is intended to burn a sample of sodium in a gas jar of chlorine, the sample is normally placed in a combustion spoon, ignited and plunged into the gas jar. It is recommended that, before ignition, a piece of dry asbestos paper be placed between the sodium and the combustion spoon.
    The main reason for this is
    A. to absorb any liquid on the surface of the sodium
    B. to absorb any moisture present in the gas
    C. to prevent the heat of the burning sodium from being conducted away by the spoon
    D. to prevent chemical attack of the spoon by the burning sodium
    E. to reduce contamination of the sodium chloride formed by chlorides produced from the metals present in the spoon

14. Bromine is in the same group of the periodic table as fluorine, chlorine, and iodine. It is
    A. insoluble in water
    B. unchanged in colour when a few drops of it are shaken up with sodium hydroxide solution
    C. less vigorous in its reaction with white phosphorus than chlorine is
    D. able to bleach moist litmus paper
    E. a brown gas at room temperature

15. A red powder is known to be the oxide of a metal. It dissolves in dilute nitric acid to form a pale-blue solution. Which of the following could it be?
    A. Copper(I) oxide
    B. Iron(III) oxide
    C. Dilead(II) lead(IV) oxide (red lead oxide)
    D. Mercury(II) oxide
    E. rust

16. From the periodic table deduce which of the following is *not* a correct statement:
    A. Element 87 would react extremely vigorously with water
    B. An advertisement speaking of a 'liquid tungsten' oil additive is misleading
    C. The element cadmium is much denser than the element rubidium
    D. Metals occur on the right-hand side of the periodic table and non-metals occur on the left-hand side
    E. Tellurium at room temperature is a solid

    (To help you find these elements on the periodic table, the relevant atomic numbers are: tungsten, 74; cadmium, 48; rubidium, 37; tellurium, 52)

17. Here are some facts about carbon. Which of them is *not* true?
    A. To investigate whether a substance contains carbon, it is heated with a sample of copper(II) oxide
    B. Carbon is contained in all living matter
    C. Carbon has a very high boiling point
    D. Carbon forms an acidic oxide characteristic of non-metals; carbon rods conduct electricity, however: a resemblance to metals
    E. Carbon is only found in compounds which occur in nature

18. The element silicon appears directly below carbon in the periodic table. Amongst the properties of silicon are:
    A. It dissolves in dilute acids to give hydrogen
    B. It does not react with alkalis
    C. It is normally obtained from its compounds by electrolysis
    D. Its compounds are common constituents of rocks
    E. It is the most abundant element in the earth's crust

19. Carbon dioxide has the formula CO_2 and silica (sand) SiO_2. The substances, however, differ widely in properties.
    The reason for this is that
    A. carbon dioxide is a gas but silica is a solid
    B. carbon dioxide dissolves slightly in water but silica is practically insoluble in water
    C. the properties of elements in the same group of the periodic table are similar but this is not true of their compounds
    D. there are exceptions to the general rule that corresponding compounds of elements in the same group have similar properties
    E. the atoms which make up carbon dioxide are arranged very differently to those which make up silica

Directions for Questions 20 to 25 (See Rubric C)

20. The periodic table was devised by Mendeleev in 1869. It is still of great importance to chemists 100 years later. Among the features of Mendeleev's table is (are):
    (i) Elements which have similar properties appear in the same period
    (ii) It is a table written in increasing order of atomic mass in such a way that similar elements occupy similar positions
    (iii) Elements with the highest boiling points are at each end of the table
    (iv) Spaces were left for elements not discovered in 1869. Mendeleev forecast the properties of these elements and, when discovered, their properties were found to be very similar to those predicted

21. The elements in Group I of the periodic table
    (i) have atomic masses differing by eight units
    (ii) are all stored under liquid paraffin
    (iii) change from non-metallic to metallic character as the atomic weight increases
    (iv) form chlorides which have similar formulae

22. In moving from Group I to Group VII across a period of the periodic table the elements
    (i) show a gradual transition in properties from metallic to non-metallic
    (ii) have oxides whose nature changes from alkaline to acidic
    (iii) have atomic numbers differing by one unit
    (iv) have atomic volumes which gradually increase

23. Imagine that a new element has been discovered which has been named Bloxhamium (symbol Bl). It has been suggested that the element might belong to the family of elements known as the alkali metals. Which of the following properties would be in accordance with this suggestion?
    (i) The element has a high melting point
    (ii) The element reacts with air so it has to be stored under water
    (iii) The element forms a chloride which is believed to have the formula $BlCl_2$
    (iv) The element forms compounds which are usually white in colour

24. Transition elements are
    (i) elements intermediate in character between metals and non-metals
    (ii) elements which form colourless compounds
    (iii) elements which can be changed into other elements by the use of nuclear reactions
    (iv) elements forming a block in the centre of the periodic table, closely related in character to each other

25. Astatine is a member of the halogen group of elements. Its atomic mass is greater than that of the other halogens, but little is known about its properties.
    It is likely that
    (i) it is a colourless liquid
    (ii) it can displace bromine from a solution of potassium bromide
    (iii) it will form a compound with sodium of formula NaAt (At being the chemical symbol for Astatine)
    (iv) it will dissolve in sodium hydroxide solution on warming

Directions for Questions 26 to 30 (See Rubric D)

Assertion		Reason
26. If an element X has a higher atomic mass than an element Y, the volume occupied by one mole of X atoms will always be greater than the volume occupied by one mole of Y atoms	BECAUSE	the atoms of an element X will always be packed together less tightly than the atoms of an element Y (assuming both are solids) if the atomic mass of X is higher than the atomic mass of Y
27. Magnesium and calcium are called 'alkali metals'	BECAUSE	the oxides of magnesium and calcium have an alkaline reaction to universal indicator

	Assertion		Reason
28.	Potassium and chlorine react vigorously together	BECAUSE	potassium reacts vigorously with water
29.	Chlorine will cause bromine to form when chlorine is bubbled through a solution of potassium bromide	BECAUSE	chlorine is a more reactive element than bromine
30.	Carbon dioxide turns lime water milky	BECAUSE	charcoal on heating with copper(II) oxide gives carbon dioxide

Chapter 14

THE ARRANGEMENT OF ATOMS IN ELEMENTS

Directions for Questions 1 to 5 (See Rubric A)

Questions 1 to 5

The following are descriptions of the structures of various substances. Select the one which applies to the substance named in the question.
A. A giant structure made up of atoms arranged tetrahedrally
B. A giant structure termed an 'a b a' structure in which the atoms in every other row are immediately above each other
C. A diatomic molecule
D. Made up of long chains of atoms
E. The atoms are arranged in a random structure

1 Magnesium

2. Plastic sulphur

3. Iodine

4. Oxygen

5. Diamond

Directions for Questions 6 to 19 (See Rubric B)

6. A Nuffield diffraction grid is a set of black dots on a piece of film. These dots give rise to a set of light spots when a bulb is viewed through the film. Two grids were looked at in this way. The light spots formed in both cases were of the same pattern but using the first grid they were closer together. Which conclusion is correct with regard to the black spots on the first piece of film compared to those on the second?
A. The spots are closer together
B. The spots are farther apart
C. The spots are fewer in number
D. The spots are larger
E. The spots are smaller

The Arrangement of Atoms in Elements

7. Which of the following processes is *not* likely to change the grain size of a metal?
   A. Alloying the metal
   B. Hammering the metal
   C. Heating the metal to red heat and then plunging it into a bath of oil
   D. Passing an electric current through the metal
   E. Rolling the metal when it is hot

8. Sulphur is an allotropic element. This means that
   A. sulphur consists of rings of 8 sulphur atoms
   B. sulphur can form chains of atoms
   C. there are several different forms of sulphur but the atoms are arranged in the same way in each form
   D. there are several different forms of sulphur because there can be different arrangements of sulphur atoms
   E. sulphur is an element which normally has a 'giant structure'

9. When powdered roll sulphur is melted and poured into a filter paper, folded as though for putting in a filter funnel, the molten sulphur on the surface solidifies. If this crust is broken off, crystals are seen. This form of sulphur is called
   A. flowers of sulphur   D. rhombic
   B. monoclinic   E. ring
   C. plastic

10. Which of the following statements is *not* true of the element carbon?
    A. It can exist in very many different structural forms
    B. One of its forms will conduct electricity
    C. It is found in all living things
    D. On burning in air, it forms an acidic gas
    E. When heated with copper(II) oxide, reduction of the oxide occurs with the formation of copper

11. The atomicity of an element is
    A. the number of atoms in a molecule of that element
    B. the number of molecules in an atom of that element
    C. the number of atoms in a mole of atoms of that element
    D. the number of atoms in a mole of molecules of that element
    E. the number of molecules in a mole of molecules of that element

12. Which of the following is a monatomic gas?
    A. Chlorine   D. Nitrogen
    B. Hydrogen   E. Oxygen
    C. Neon

13. Which of the following contains the greatest number of molecules, given that an atom of chlorine is almost twice as heavy as an atom of fluorine?
    A. 10 g of CF_4
    B. 10 g of CF_3Cl
    C. 10 g of CF_2Cl_2
    D. 10 g of $CFCl_3$
    E. 10 g of CCl_4

14. The number of grammes of dinitrogen monoxide N_2O, having the same number of molecules as 2 moles (92 g) of nitrogen dioxide molecules, NO_2, is
    A. 2   B. 22   C. 44   D. 88   E. 92

15. The analysis of a certain substance showed that 7 g of nitrogen were combined with 1 g of hydrogen. The simplest formula of the substance must therefore be
    A. NH_2 B. NH_7 C. N_2H D. N_2H_4 E. N_7H

16. White phosphorus consists of P_4 molecules. The mass of $\frac{1}{4}$ mole of white phosphorus molecules, P_4 is
    A. 0.25 g   B. 4 g   C. 7.75 g   D. 31   E. 124

17. A chemist needed 0.1 mole of anhydrous copper sulphate, $CuSO_4$. He only possessed blue copper sulphate crystals, however. What would be the minimum mass of these crystals, having the formula $CuSO_4, 5H_2O$, that he would need in order to produce the required amount of anhydrous copper sulphate?
    A. 15.8 g   B. 17.9 g   C. 24.8 g   D. 44.8 g   E. 53.8 g

18. 8 g of hydrogen would occupy 96 dm³ at room temperature. What mass of helium would occupy the same volume under the same conditions?
    A. 2 g   B. 4 g   C. 8 g   D. 16 g   E. 32 g

19. A certain element forms a gaseous compound with hydrogen (formula XH_4). At room temperature, 48 dm³ of the compound weigh 64 g. Which of the following is the atomic mass of the element X?
    A. 7   B. 8   C. 28   D. 32   E. 60

Directions for Questions 20 to 26 (See Rubric C)

20. The atoms in a piece of metallic lead
    (i) can be seen on photographs taken by X-rays
    (ii) are arranged in a regular pattern
    (iii) form crystals consisting of small definite numbers of atoms
    (iv) are packed together as closely as possible

The Arrangement of Atoms in Elements

21. Graphite has a giant structure made up from layers of atoms; the atoms in each layer being arranged in hexagons. The distance between the layers is considerably greater than the distance between the atoms. It follows from this that
    (i) graphite has much weaker forces between the layers than between the atoms in each layer
    (ii) a crystal of graphite is very easy to cleave in one direction
    (iii) graphite has a slippery feel
    (iv) graphite has a very high boiling point

22. Which of these substances has a cubic structure for the arrangement of its constituent particles?
    (i) Common salt
    (ii) Copper
    (iii) Magnesium oxide
    (iv) Water

23. The chemical formula of a substance can be used to determine
    (i) the mass of each element in a given mass of the substance
    (ii) the volume of a given mass of the substance, should it be a gas
    (iii) the proportions by mass of the elements in the substance
    (iv) the type of structure possessed by the substances

24. One mole of molecules of any gas
    (i) contains the Avogadro constant of molecules
    (ii) is twice the mass of one mole of atoms of the same gas
    (iii) contains as many molecules as there are atoms in one mole of hydrogen atoms, H
    (iv) occupies about 24 cm^3 at laboratory temperature and atmospheric pressure

25. The atomic mass of oxygen is 16 and the atomic mass of hydrogen is 1. It follows that
    (i) a mole of oxygen atoms, O, is sixteen times as heavy as a mole of hydrogen atoms, H
    (ii) a mole of oxygen molecules, O_2, is sixteen times as heavy as a mole of hydrogen molecules, H_2
    (iii) 16 g of oxygen react exactly with 2 g of hydrogen when water is produced
    (iv) oxygen is sixteen times as dense as hydrogen

26. When sulphur is burned in oxygen to give sulphur dioxide, it is found that the volume of sulphur dioxide formed is exactly the same as the volume of oxygen used. This suggests that the molecule of sulphur dioxide contains the same number of oxygen atoms as does the oxygen molecule. On this evidence *alone*, which of the following formulae is (are) possible for the molecule of sulphur dioxide?
    (i) SO   (ii) SO_2 (iii) SO_3 (iv) S_2O_2

Directions for Questions 27 to 35 (See Rubric D)

	Assertion		Reason
27.	X-ray diffraction is used for investigating the structure of crystals	BECAUSE	X-rays are deflected by the different layers of atoms in a crystal
28.	A metal sample made from large crystals is stronger than one made from small crystals	BECAUSE	large crystals are held together more firmly than small ones
29.	When burnt in excess of oxygen, equal masses of graphite and diamond give the same volume of carbon dioxide	BECAUSE	graphite and diamond are both pure forms of the element carbon
30.	Sulphur, on heating to just above its melting point is runny. On further heating, the sulphur becomes very viscous	BECAUSE	when sulphur melts, it forms interlocking rings
31.	Rhombic sulphur is crystalline	BECAUSE	the molecules in rhombic sulphur contain eight atoms
32.	Iodine vaporizes more easily than graphite	BECAUSE	iodine and graphite are in different groups of the periodic table
33.	The mass of a mole of sulphur molecules is eight times the mass of a mole of sulphur atoms	BECAUSE	in a molecule of sulphur, the atoms are arranged in a ring
34.	Gaseous oxygen and liquid oxygen are allotropes	BECAUSE	allotropes are different forms of the same element
35.	The formula for hydrogen chloride gas is HCl	BECAUSE	hydrogen chloride gas contains the same number of molecules as an equal volume of hydrogen, both gases being under the same conditions

Chapter 15

SOLIDS, LIQUIDS, AND GASES

Directions for Questions 1 to 3 (See Rubric A)

Questions 1 to 3

Below are tabulated some properties of the chlorides of five metals A–E:

	Colour	Melting point, °C	Boiling point, °C
A.	White	714	1418
B.	Brown	498	993
C.	Colourless	−33	113
D.	Yellow	1001	987
E.	Blue	724	1050

Which of these chlorides

1. is liquid at room temperature?
2. can be vaporized completely before being melted?
3. is most likely to have the greatest value for the heat of vaporization?

Directions for Questions 4 to 9 (See Rubric B)

4. In Mexico City water normally boils at 93 °C. This is because
   A. the water in Mexico City is not very pure
   B. the city is situated at a high altitude
   C. the boiling point of the water varies with the temperature of the surroundings
   D. the water is very soft
   E. the water is heavily treated with fluoride

5. Which of the following statements is true for a substance with a giant structure?
   A. It has a low heat of vaporization
   B. It has a low melting point
   C. It has a low density
   D. It has particles which, in the solid state, are regularly arranged
   E. It is made up of particles which are held together by weak forces

6. A Bunsen burner was found to transmit heat energy at the rate of 3088 J min^{-1}. It was used to heat some water in a flask. After boiling for 10 mins, it was found that 13.65 g of water had evaporated. The heat of vaporization of water molecules is equal to

A. $\dfrac{10 \times 3088}{13.65}$ kJ mol^{-1}

B. $\dfrac{10 \times 3088}{13.65 \times 100}$ kJ mol^{-1}

C. $\dfrac{10 \times 3088 \times 1000}{13.65}$ kJ mol^{-1}

D. $\dfrac{10 \times 3088 \times 18}{13.65 \times 100}$ kJ mol^{-1}

E. $\dfrac{10 \times 3088 \times 18}{13.65 \times 1000}$ kJ mol^{-1}

7. The heat of vaporization of trichloromethane is 40.00 kJ mol^{-1} of trichloromethane molecules CHCl$_3$, at its boiling point and that of tetrachloromethane is 45.40 kJ mol^{-1} of tetrachloromethane molecules, CCl$_4$, at its boiling point. A mole of trichloromethane molecules, CHCl$_3$, mass 119.5 g and a mole of tetrachloromethane molecules, CCl$_4$, mass 154 g.

The ratio of the amount of heat necessary to vaporize 1 g of tetrachloromethane to the amount of heat necessary to vaporize 1 g of trichloromethane, is therefore

A. $\dfrac{40.00}{45.40}$

B. $\dfrac{45.40}{40.00}$

C. $\dfrac{40.00}{45.40} \times \dfrac{154}{119.5}$

D. $\dfrac{45.40}{40.00} \times \dfrac{154}{119.5}$

E. $\dfrac{45.40}{40.00} \times \dfrac{119.5}{154}$

8. A source of heat supplying 1200 J min^{-1} brings 100 g of liquid X (specific heat capacity 3 kJ kg^{-1} °C^{-1}) from 20.0 °C to its boiling point in exactly fifteen minutes. Assuming no loss of heat to the surroundings, liquid X therefore boils at
A. 40 °C B. 60 °C C. 80 °C D. 100 °C
E. It is impossible to say from the information given

Solids, Liquids, and Gases

9. The heat required to raise the temperature of 1 mole of atoms of an element by 1 °C is approximately the same for most elements.

    The heat required to raise the temperature of 207 g of lead by 1 °C is 26.82 J. The heat required to raise the temperature of 1 g of a certain other metallic element by 1 °C is 0.65 J.

    Which of the following procedures will give the approximate atomic mass of the latter element?

    A. $26.82 \div 0.65$
    B. 26.82×0.65
    C. $207 \div 26.82$
    D. 207×26.82
    E. $\dfrac{207 \times 0.65}{26.82}$

Directions for Questions 10 to 12 (See Rubric C)

10. An investigation is made into the cooling of benzene from room temperature down to its freezing point (5 °C). It is found that
    (i) all the benzene suddenly solidifies in one mass when the freezing point is reached
    (ii) the benzene cools faster as the temperature gets nearer to the freezing point
    (iii) when the benzene freezes, heat is absorbed from the surroundings
    (iv) if a graph of temperature against time is plotted during the cooling, a definite break occurs in the shape of the graph when the freezing point is reached

11. The temperature at which pure alcohol boils
    (i) gradually increases as the alcohol boils away
    (ii) varies with the temperature of the room
    (iii) is lowered by heating the alcohol with an immersion heater
    (iv) falls slightly if the pressure of the atmosphere falls

12. Brownian motion is the name given to
    (i) the movement of molecules more commonly known as diffusion
    (ii) the movement of very small particles when placed in a liquid, due to bombardment of the particles by the molecules of the liquid
    (iii) the movement by which it is possible for molecules with sufficient energy to escape from the surface of a heated liquid
    (iv) the movement of a smoke particle in a smoke cell

Directions for Questions 13 to 15 (See Rubric D)

	Assertion		Reason
13.	When a liquid boils, more particles are escaping from the surface of the liquid than are returning to it	BECAUSE	only when a liquid boils is sufficient energy being applied to a particle to enable it to escape from the surface of the liquid
14.	Water has a surprisingly high heat of vaporization per mole of water molecules, H_2O. This would indicate the existence of forces between molecules of water of a type different to those between molecules of, say, trichloromethane	BECAUSE	the mass of a mole of trichloromethane molecules, $CHCl_3$ (119), is more than six times greater than that of water (18), but water has a higher heat of vaporization per mole of molecules
15.	Sodium is more easily vaporized than lithium	BECAUSE	the forces between sodium atoms are weaker than those between lithium atoms

Chapter 16

ELECTROLYSIS

Directions for Questions 1 to 11 (See Rubric A)

Questions 1 to 6

Five possible arrangements of molecules and ions are given below:
A. closely packed ions
B. widely separated ions
C. widely separated molecules
D. closely packed molecules with a random arrangement
E. closely packed molecules with a regular arrangement

Which is most likely to describe the arrangement of particles in the substances in italics?

1. Gaseous *ammonia*

2. *Sodium chloride* when in a dilute aqueous solution

3. Liquid *bromine*

4. Crystals of *potassium chloride*

5. Solid *iodine*

6. 0.01 g of *iodine* dissolved in 1 dm³ of benzene

Questions 7 to 11

Which of the five values below applies to the charge on the ion in italics in each of the following questions?
A. +1   D. −1
B. +2   E. −2
C. +3

7. *One chloride ion* in aluminium chloride, AlCl₃. (It is known that 3 moles of electrons liberate 1 mole of aluminium atoms, Al, during electrolysis.)

8. *The zinc ion* in zinc phosphate, $Zn_3(PO_4)_2$. (It is known that the phosphate ion carries three charges.)

9. *The cadmium ion* in cadmium sulphide, CdS. (It is known that the formula for sodium sulphide is Na_2S and that the sodium ion carries a single positive charge)

10. *The carbonate ion* in sodium carbonate $(Na^+)_2CO_3^x$

11. *The copper ion* in red copper oxide, Cu_2O. (It is known that when red copper oxide changes into black copper oxide, each copper atom loses an electron; in black copper oxide the copper has a double positive charge)

Directions for Questions 12 to 22 (See Rubric B)

12. When molten lead iodide (formula PbI_2) is electrolysed, iodine is formed at the anode and lead is formed at the cathode. Solid lead iodide does not conduct electricity. Which of the following is *not* a reasonable deduction?
    A. The current is being carried by charged particles
    B. Since the lead iodide itself is neutral, there must be equal numbers of positive and negative charges present in the substance
    C. Lead ions are positively charged and iodide ions are negatively charged
    D. The number of charges on lead ions must be twice that on iodide ions
    E. There are no ions in solid lead iodide

13. In the electrolysis of magnesium chloride two moles of chlorine atoms, Cl, are formed for each mole of magnesium atoms, Mg. Which of the following is the most likely explanation of this?
    A. The atomic mass of chlorine is twice that of magnesium
    B. The charge on a magnesium ion is twice that on a chloride ion
    C. The charge on a chloride ion is twice that on a magnesium ion
    D. The size of a chloride ion is twice that of a magnesium ion
    E. The formula for magnesium chloride is Mg_2Cl

Electrolysis

14. Which of the circuits in Figure 16.1 would be the most suitable for investigating how much lead is produced during the electrolysis of molten lead(II) iodide?

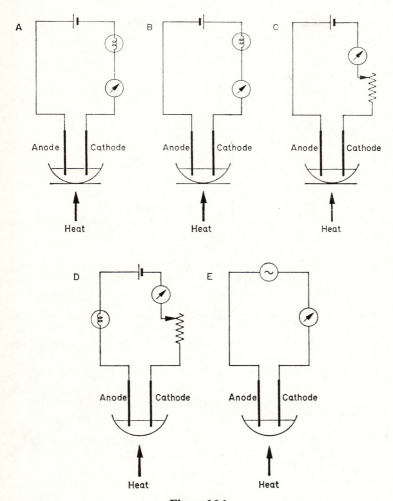

Figure 16.1

15. It was found that when a certain molten lead salt was electrolysed a current of 0.6 A passed for 90 min produced 6.9 g of lead. The amount of electricity needed to produce 1 mole of lead atoms, Pb, is therefore

   A. $\dfrac{207}{6.9} \times 90 \times 60 \times 0.6$ coulombs

   B. $\dfrac{207}{6.9} \times 90 \times 0.6$ coulombs

   C. $\dfrac{207}{6.9} \times 90 \times 0.60 \times 6$ faradays

   D. $\dfrac{207}{6.9} \times 90 \times 0.6$ faradays

   E. $\dfrac{6.9}{207} \times 90 \times 60 \times 0.6$ faradays

16. The electrolysis of a certain liquid resulted in the formation of hydrogen at the cathode and a gas which turned moist starch-iodide paper blue at the anode. Which of the following could it be?
    A. A solution of copper chloride in water
    B. A solution of sodium chloride in water
    C. A solution of sulphuric acid in water
    D. Molten sodium chloride
    E. Alcohol

17. A mole of electrons is the amount of electricity required to produce
    A. an atom of silver
    B. an Avogadro constant of atoms of silver, from a solution containing $Ag^+(aq)$ ions
    C. a mole of lead atoms, Pb, from a solution containing $Pb^{2+}(aq)$ ions
    D. 2 g of hydrogen (the mass of a mole of hydrogen is 2 g)
    E. a molecule of hydrogen

18. During electrolysis of a hot aqueous solution containing only sodium chloride and sodium hydroxide, using copper electrodes, it is found that
    A. sodium is produced at the cathode
    B. copper is transferred from the anode to the cathode
    C. a reddish-orange precipitate of copper(I) oxide is formed
    D. the passage of 4 moles of electrons of electricity would remove 2 moles of copper atoms, Cu, from the anode
    E. the passage of an equivalent amount of electricity through an electrolytic cell containing copper sulphate solution, using copper electrodes, would result in the same changes in mass occurring in both sets of electrodes

Electrolysis

19. Solutions of copper sulphate (CuSO$_4$) and silver nitrate (AgNO$_3$) are electrolysed by identical currents flowing for the same length of time, using electrodes made of carbon. It was found that the number of moles of copper atoms (Cu) formed
  A. on the anode by the electrolysis of the copper sulphate solution was equal to the number of moles of silver atoms, Ag, formed on the anode by the electrolysis of the silver nitrate
  B. on the anode by the electrolysis of the copper sulphate solution was equal to half the number of moles of silver atoms, Ag, formed on the anode by the electrolysis of the silver nitrate
  C. on the anode by the electrolysis of the copper sulphate solution was equal to twice the number of moles of silver atoms, Ag, formed on the anode by the electrolysis of the silver nitrate
  D. on the cathode by the electrolysis of the copper sulphate solution was equal to half the number of moles of silver atoms, Ag, formed on the cathode by the electrolysis of the silver nitrate
  E. on the cathode by the electrolysis of the copper sulphate solution was equal to twice the number of moles of silver atoms, Ag, formed on the cathode by the electrolysis of the silver nitrate

20. A strip of damp filter paper is laid on a microscope slide and held to the slide at the ends by crocodile clips connected to a battery. A spot of lead nitrate solution and a spot of potassium iodide solution are placed in the positions shown in Figure 16.2.

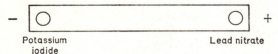

Potassium iodide   Lead nitrate

Figure 16.2

Which of the following would you expect to observe first?
  A. A streak of iodine near the crocodile clip connected to the negative of the battery
  B. A streak of iodine near the crocodile clip connected to the positive of the battery
  C. A streak of lead iodide near the centre of the slide
  D. The appearance of iodine near the centre of the slide
  E. No sign of any colour

21. A large number of substances contain ions. Which of the following is *not* a correct statement about such particles?
  A. They are always much less stable than the atoms to which they are related
  B. They often have chemical properties which are very different to those of their related atoms
  C. Those ions which travel towards the negative electrode are called cations; those which travel to the positive electrode are called anions
  D. In solid electrolytes an electrostatic force of attraction holds the ions together in a lattice
  E. The ion discharged at the cathode when an aqueous solution of a substance is electrolysed may not be the same as the ion discharged when the same substance is electrolysed in the molten state

22. The atoms in a molecule are held together by
  A. weak forces between the atoms
  B. an electrostatic force of attraction between charged atoms
  C. an electrostatic force of attraction between charged ions
  D. a force of attraction between the nucleus of one atom and the electrons of another
  E. a pair of electrons shared between the atoms

Directions for Questions 23 to 29 (See Rubric C)

23. When an ion is formed from an atom
  (i) the nucleus of the atom is unchanged
  (ii) it has the same colour as the atom
  (iii) it either gains or loses one, two, or three electrons
  (iv) the reverse process can only be brought about by electrolysis

24. The electrolysis of copper sulphate solution using copper electrodes results in
  (i) the anode dissolving
  (ii) the formation of oxygen at one electrode
  (iii) no change in the concentration of copper ions in solution
  (iv) a gradual increase in the acidity of the solution

25. An aqueous solution of silver nitrate yields silver on electrolysis. The amount of silver deposited is increased by
  (i) increasing the concentration of the solution
  (ii) increasing the depth to which the electrodes were immersed in the solution
  (iii) increasing the time for which the current was passed
  (iv) increasing the separation of the electrodes

Electrolysis

26. When a metal is deposited from a solution of one of its salts during electrolysis, the mass of metal deposited by a given quantity of electricity depends on
    (i) the current passing through the solution
    (ii) the charge on the metal ion
    (iii) the voltage used
    (iv) the atomic mass of the metal

27. The quantity of electricity needed to produce 4 moles of zinc atoms, Zn, from $Zn^{2+}(aq)$ ions yields from a solution containing $H^+(aq)$ ions:
    (i) 4 moles of hydrogen molecules, H_2
    (ii) 4 moles of hydrogen atoms, H
    (iii) 8 moles of hydrogen atoms, H
    (iv) 8 moles of hydrogen molecules, H_2

28. An Avogadro constant of particles is contained in
    (i) a faraday
    (ii) a coulomb
    (iii) a mole of electrons
    (iv) an atom

29. Electrolytes can be distinguished from non-electrolytes because non-electrolytes
    (i) alone do not conduct a current of electricity when solid
    (ii) all have very low melting points
    (iii) do not form crystals
    (iv) are not decomposed by electricity when in aqueous solution

Directions for Questions 30 to 35 (See Rubric D)

Assertion		Reason
30. When lead bromide is melted an electric current can pass through it; if a bulb is included in the circuit, it will light up	BECAUSE	metals like lead are normally good conductors of electricity
31. Before electrolysis, the ions in molten lead bromide are stationary	BECAUSE	ions move to the electrodes only when a current is passed

Assertion		Reason
32. Ions of the same element always carry the same charge	BECAUSE	atoms of the same element always have the same number of negative charges surrounding the positively charged nucleus
33. Solid sodium chloride conducts electricity	BECAUSE	solid sodium chloride is composed of ions
34. During the electrolysis of dilute sulphuric acid, the quantities of electricity exchanged at the cathode and anode respectively are equal	BECAUSE	the products, hydrogen gas and oxygen gas, are both composed of diatomic molecules
35. If a crystal of potassium permanganate is placed on a damp filter paper resting on a glass slide and the paper is connected by clips to a battery, the purple colour is seen to move towards the anode	BECAUSE	the electrode connected to the positive pole of the battery is called the anode and attracts ions of the opposite charge towards it

Chapter 17

REACTING QUANTITIES

Directions for Questions 1 to 5 (See Rubric A)

Questions 1 to 5

    A. 2   B. 4   C. 12   D. 24   E. 72

Choose the answers to the following questions from the above list of possible answers:

1. The approximate volume in cubic decimetres occupied by 1 mole of oxygen atoms, O, at room temperature and pressure

2. The number of moles of hydrogen ions, $H^+(aq)$, in 1 dm^3 of 2 M sulphuric acid (H_2SO_4).

3. The volume in cubic centimetres of 0.5 M sodium hydroxide which reacts with 12 cm^3 of 1 M iron(III) chloride solution in the reaction represented by

$$Fe^{3+}(aq) + 3OH^-(aq) \rightarrow Fe(OH)_3(s)$$

4. The mass of pentane (C_5H_{12}) produced from 33 g of iodopentane ($C_5H_{11}I$) in the reaction represented by

$$C_5H_{11}I(l) + H_2(g) \rightarrow C_5H_{12}(l) + HI(g)$$

5. The maximum volume in cubic centimetres of oxygen gas that can be produced at room temperature and pressure from 10 cm^3 of a 0.1 M solution of hydrogen peroxide which decomposes according to the equation:

$$2H_2O_2(aq) \rightarrow 2H_2O(l) + O_2(g)$$

Directions for Questions 6 to 13 (See Rubric B)

6. If you wished to make 250 cm^3 of a 0.02 M solution of potassium permanganate, $KMnO_4$, which one of the following masses of potassium permanganate would you use?
    A. 0.316 g   B. 0.79 g   C. 1.58 g   D. 3.16 g   E. 7.9 g

7. The reaction between tin(II) ions and iron(III) ions may be represented by
$$Sn^{2+}(aq) + 2Fe^{3+}(aq) \rightarrow Sn^{4+}(aq) + 2Fe^{2+}(aq)$$
1 M tin(II) chloride is added to 40 cm³ of 2 M iron(III) chloride until all the iron(III) ions have been converted to iron(II) ions. How much of the tin(II) chloride solution is needed?
A. 10 cm³ B. 20 cm³ C. 40 cm³ D. 80 cm³ E. 160 cm³

8. When the dropper in Figure 17.1 is squeezed, the solution of silver nitrate falls into the solution of sodium chloride and the mixture becomes white and thick as a precipitate of silver chloride is formed.

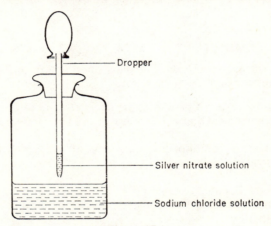

Figure 17.1

What happens to the total mass of the dropping bottle and its contents?
A. It increases in mass due to the formation of the heavy precipitate
B. It remains the same because only a physical change takes place
C. It decreases in mass because more air enters the dropping tube
D. Sometimes it increases in mass and sometimes decreases
E. The mass remains unchanged throughout the experiment

9. On adding a solution of silver nitrate to a solution of a chloride, a white precipitate is produced. A bottle was labelled Indium Chloride Solution (0.1 M) but it did not give the formula for indium chloride. A student decided to add varying amounts of (0.2 M) silver nitrate solution to 6 cm³ of the indium chloride solution. Each time he centrifuged the mixture for two minutes before measuring the height of the precipitate. He then plotted the graph shown in Figure 17.2.

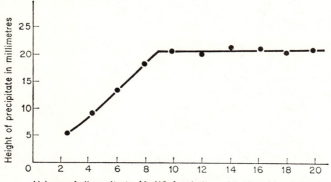

Figure 17.2

Silver ions react with the chloride ions thus:
$$Ag^+(aq) + Cl^-(aq) \rightarrow AgCl(s)$$
Which of the following is the formula for indium chloride?
A. $InCl_2$ B. $InCl_3$ C. In_2Cl_3 D. In_3Cl_2 E. In_3Cl

10. When hydrogen chloride is passed over iron wool, iron chloride and hydrogen are produced. 1.4 g of iron produce 600 cm³ of hydrogen at room temperature.
   Which of the following best represents the equation for this reaction?
A. $Fe + HCl \rightarrow FeCl + H$
B. $Fe + 2HCl \rightarrow FeCl_2 + H_2$
C. $Fe + 4HCl \rightarrow FeCl_4 + 2H_2$
D. $2Fe + 2HCl \rightarrow 2FeCl + H_2$
E. $2Fe + 6HCl \rightarrow 2FeCl_3 + 3H_2$

11. It is found that 1.35 g of aluminium combines with sulphur to give 3.75 g of aluminium sulphide.
   The equation for the reaction is therefore
A. $Al + 3S \rightarrow AlS_3$
B. $2Al + 3S \rightarrow Al_2S_3$
C. $3Al + 2S \rightarrow Al_3S_2$
D. $4Al + S_8 \rightarrow 2AlS_2$
E. $5Al + 2S_8 \rightarrow Al_5S_{16}$

12. When potassium carbonate reacted with excess 2 M hydrochloric acid 0.004 moles of potassium carbonate yielded 96 cm^3 of carbon dioxide gas. It is also known that 10 cm^3 of the same acid reacted with 10 cm^3 of 1 M potassium carbonate solution.

    The equation must therefore include the following terms:
    A. xCO$_3{}^{2-}$(aq), xH$^+$(aq) and xCO$_2$(g)
    B. xCO$_3{}^{2-}$(aq), $2x$H$^+$(aq) and xCO$_2$(g)
    C. xCO$_3{}^{2-}$(aq), $2x$H$^+$(aq) and $2x$CO$_2$(g)
    D. $2x$CO$_3{}^{2-}$(aq), xH$^+$(aq) and xCO$_2$(g)
    E. $2x$CO$_3{}^{2-}$(aq), xH$^+$(aq) and $2x$CO$_2$(g)

13. A syringe was filled with 50 cm^3 of bromine vapour. Another syringe was filled with 40 cm^3 of hydrogen sulphide gas. The two syringes were connected by a three-way tap and the tap was then opened to allow the gases to mix.

    A reaction took place in accordance with the equation:
    $$Br_2(g) + H_2S(g) \rightarrow 2HBr(g) + S(s)$$
    The volume of gas remaining at the end of the reaction was
    A. 10 cm^3 B. 80 cm^3 C. 90 cm^3 D. 135 cm^3 E. 180 cm^3

Directions for Questions 14 to 19 (See Rubric C)

14. A chemical equation
    (i) can be written only in one way for a particular reaction
    (ii) can be written only when the exact conditions under which the reaction is carried out are stated
    (iii) for an ionic reaction can always be predicted if the charges on the reacting ions are known
    (iv) should be written only when an experiment has been carried out to determine the relative numbers of moles of particles which react together and are produced

15. An equation for the reaction between sodium hydroxide and sulphuric acid solution has been found by experiment to be:
    $$2NaOH(aq) + H_2SO_4(aq) \rightarrow Na_2SO_4(aq) + 2H_2O(l)$$
    10 cm^3 of 0.5 M sulphuric acid is used and the amount of sodium hydroxide solution needed to complete the reaction and leave no excess is
    (i) 40 cm^3 of 0.25 M sodium hydroxide
    (ii) 20 cm^3 of 0.5 M sodium hydroxide
    (iii) 10 cm^3 of 1 M sodium hydroxide
    (iv) 2.5 cm^3 of 2 M sodium hydroxide

Reacting Quantities

16. 1 dm³ of 1 M oxalic acid reacts with 2 dm³ of 1 M potassium hydroxide. From this it follows that
    (i) 2 dm³ of 0.5 M oxalic acid react with 1 dm³ of 2 M potassium hydroxide
    (ii) 0.5 dm³ of 2 M oxalic acid react with 1 dm³ of 1 M potassium hydroxide
    (iii) 0.5 dm³ of 2 M oxalic acid react with 4 dm³ of 0.5 M potassium hydroxide
    (iv) 0.5 dm³ of 0.5 M oxalic acid react with 0.25 dm³ of 1 M potassium hydroxide

17. When 0.1 mole of atoms of a metallic element X was added to excess of an aqueous solution of a copper(II) salt, it all dissolved and 0.1 mole of atoms of copper was formed in its place. On this evidence it appears that
    (i) X is more reactive than copper
    (ii) the ions of X formed in solution each had two positive charges
    (iii) all the atoms of X had been changed into ions
    (iv) all the copper ions had been changed into atoms

18. When iron filings are reacted with copper sulphate solution the copper powder which is precipitated is washed first with water and then with acetone. Acetone is used because
    (i) it burns easily
    (ii) it is a good solvent
    (iii) it is insoluble in water
    (iv) it evaporates very readily

19. The reaction between magnesium ribbon and dilute hydrochloric acid takes place in accordance with the equation
$$Mg(s) + 2HCl(aq) \rightarrow MgCl_2(aq) + H_2(g)$$
Information that can be derived from this equation using *only* a table of atomic masses includes
    (i) 2.4 g of magnesium would need 200 cm³ of 1 M HCl to react completely
    (ii) 2.4 g of magnesium would give rise to 240 cm³ of hydrogen measured under room conditions
    (iii) In order to produce 0.4 g of hydrogen it would be necessary to use a solution containing 28.8 g of hydrogen chloride
    (iv) This reaction takes place at room temperature

Directions for Questions 20 to 25 (See Rubric D)

	Assertion		Reason

Assertion   **Reason**

20. A solution of sulphuric acid H_2SO_4, containing 4.9 g of sulphuric acid in 100 cm³ of water is a 0.05 M solution   BECAUSE   a 1 M solution of a substance is 1 M with respect to each of the ions present in the substance

21. In a molar aqueous solution of iron(III) chloride ($FeCl_3$) there is 1 mole of $Fe^{3+}(aq)$ ions and 1 mole of $Cl^-(aq)$ ions   BECAUSE   a molar solution of a substance is molar with respect to each of the ions present in the substance

22. The reaction between aqueous solutions of lead nitrate ($Pb(NO_3)_2$) and potassium iodide (KI) is an example of an 'ionic association' reaction   BECAUSE   the reaction occurring between lead ions and iodide ions, both in aqueous solution, is $Pb^{2+}(aq) + 2I^-(aq) \rightarrow PbI_2(s)$

23. When investigating the equation for the reaction between iron filings and copper(II) sulphate solution, excess iron is used and the reaction is regarded as complete when the copper sulphate solution becomes colourless   BECAUSE   a solution of copper sulphate is blue: when all the copper has been displaced from such a solution, the blue colour is lost

24. 64 g of sulphur dioxide (SO_2) occupy a volume twice as large as 32 g of oxygen under the same conditions   BECAUSE   masses of gases may be related directly to their volume by a knowledge of the volume occupied by one mole of gas molecules under the conditions of the experiment

Assertion		**Reason**
25. Before attempting to measure the volume of gas produced when dilute hydrochloric acid is added to a known mass of sodium carbonate, the hydrochloric acid to be used is treated with a few spatulas of sodium carbonate	BECAUSE	failure to add a few spatulas of sodium carbonate to the acid before use results in less gas than expected being collected when dilute hydrochloric acid is added to a known mass of sodium carbonate

Chapter 18

RATES OF REACTION

Directions for Questions 1 to 8 (See Rubric A)

Questions 1 to 4

When dilute hydrochloric acid is added to a solution of sodium thiosulphate, a milky precipitate of sulphur is formed. The initial rate of reaction may change in any of these ways:
A. The reaction becomes instantaneous
B. The rate of reaction is decreased
C. The rate of reaction is increased
D. The rate of reaction is approximately doubled
E. The rate of reaction is unchanged

Choose from these five effects the one which is appropriate to each of the following situations (assuming that the concentration of the sodium thiosulphate solution is unaltered):

1. The concentration of acid used is decreased

2. The volume of acid used is increased

3. The solution of sodium thiosulphate is made from a finely ground powder instead of crystals

4. The temperature is raised from 20 °C to 30 °C

Rates of Reaction

Questions 5 to 8

(In the questions below assume that the first-named is plotted along the *y*-axis.)

Which of the graphs in Figure 18.1 might represent

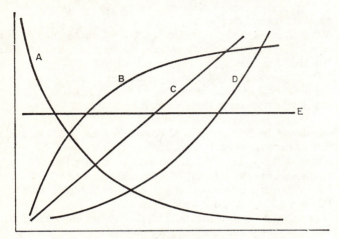

Figure 18.1

5. the mass of a catalyst at different stages of a reaction plotted against time?

6. the volume of carbon dioxide produced by reaction between calcium carbonate and dilute hydrochloric acid plotted against the time since the two substances were first mixed?

7. a plot of the time taken to produce a visible precipitate against the concentration of sodium thiosulphate used in a series of reactions when a constant volume of 2 M hydrochloric acid is added to a constant volume of sodium thiosulphate solutions differing in concentration?

8. a plot of rate of reaction against concentration of sodium thiosulphate used for a series of reactions in which a constant volume of 1 M hydrochloric acid is added to a constant volume of sodium thiosulphate solutions differing in concentration?

Directions for Questions 9 to 13 (See Rubric B)

9. Which of the following statements is *not* likely to be correct?
   A. Coal dust can be oxidized explosively
   B. In many reactions a catalyst is used in a finely divided state
   C. In the thermit reaction aluminium powder is used in preference to aluminium turnings
   D. When lead tartrate is heated lead is produced in a very finely divided state. This form of lead is very much more reactive towards air than is, for example, lead foil
   E. When manganese(IV) oxide is added to 10 cm^3 of a 20-volume solution of hydrogen peroxide, more oxygen is obtained when it is finely powdered than when it is in granular form

10. When dilute hydrochloric acid (concentration 2 M) is added to sodium carbonate crystals, carbon dioxide is evolved. Which of the following will *not* speed up the initial rate of reaction?
    A. Using double the amount of hydrochloric acid
    B. Increasing the concentration of the hydrochloric acid to 4 M, but using the same quantity of acid
    C. Increasing the concentration of the hydrochloric acid to 4 M but using only half the quantity of acid
    D. Grinding the crystals of sodium carbonate into a powder
    E. Carrying out the reaction at a higher temperature

11. A catalyst may best be defined as
    A. a substance which speeds up a chemical reaction
    B. a substance which changes the rate of a chemical reaction
    C. a substance which changes the rate of a chemical reaction without itself taking part in the chemical reaction
    D. a substance which changes the rate of a chemical reaction but is present in the same amount at the end of the reaction as at the beginning
    E. a substance which it is necessary to add in order to make a chemical reaction take place

12. Potassium chlorate is heated until it just starts to produce oxygen. It is then allowed to cool a little until all trace of oxygen evolution ceases. What happens if manganese(IV) oxide is then added?
    A. Nothing
    B. The manganese(IV) oxide liberates its oxygen
    C. Chemical interaction takes place to form a new substance
    D. Oxygen is produced when the mixture is reheated
    E. Evolution of oxygen is quite vigorous

Rates of Reaction

Directions for Questions 13 to 16 (See Rubric C)

13. When a tuft of dry platinized asbestos is placed in the mouth of a gas jar of hydrogen in the laboratory
    (i) the hydrogen burns quietly
    (ii) the platinized asbestos glows red
    (iii) nothing happens unless the platinized asbestos is heated
    (iv) the gas explodes and water is formed

14. The decomposition of hydrogen peroxide is accelerated by
    (i) adding manganese(IV) oxide
    (ii) raising the temperature
    (iii) adding a suitable enzyme
    (iv) adding water

15. When the decomposition of hydrogen peroxide by means of a catalyst is being investigated, it is found that
    (i) hydrogen gas is evolved
    (ii) a different gas is evolved if a catalyst is not used
    (iii) the quantity of catalyst used has no effect on the reaction rate
    (iv) blood will also exert a catalytic effect upon hydrogen peroxide

16. The rate of a reaction normally decreases as the reaction gets nearer to completion. Reasons for this normally include:
    (i) Energy is taken from the mixture in order to form new substances; the temperature of the mixture therefore decreases and the reaction becomes slower
    (ii) In the case of reactions between a solid and a liquid, the particles of the reacting solid coalesce together as the reaction proceeds thus reducing their surface area
    (iii) The product acts as a negative catalyst upon the reaction
    (iv) The reacting substances are present in less concentrated form

Directions for Questions 17 to 20 (See Rubric D)

Assertion		Reason
17. The energy needed to break bonds and to form other bonds during a chemical reaction is supplied more readily when the temperature is raised	BECAUSE	on carrying out a reaction at a high temperature the number of collisions between the reacting particles is increased
18. Catalysts are used in many industrial reactions	BECAUSE	catalysts do not change the products of a reaction—their role is normally to obtain the products more quickly

Assertion		**Reason**
19. Potassium chlorate gives off oxygen more easily when it is heated with copper(II) oxide than when it is heated alone	BECAUSE	both copper(II) oxide and potassium chlorate give off oxygen on heating
20. The physical state of a catalyst always remains unchanged during a reaction	BECAUSE	a catalyst, since it is not consumed in a reaction, cannot be involved directly in the reaction it is catalysing

Chapter 19

EQUILIBRIA

Directions for Questions 1 to 3 (See Rubric A)

Questions 1 to 3

Iodine dissolves in both potassium iodide and trichloromethane. Possible colour changes occurring when these solutions are used include:
A. The brown colour becomes darker
B. The brown colour becomes lighter
C. The purple colour becomes darker
D. The purple colour becomes lighter
E. The colourless solution becomes brown

1. What is the colour change occurring in a solution of iodine in potassium iodide when the solution is shaken with trichloromethane?

2. What is the colour change occurring in a solution of iodine in trichloromethane when the solution is shaken with some potassium iodide solution?

3. After shaking a solution of iodine in trichloromethane with some potassium iodide solution, the latter is discarded. What is the colour change occurring in the potassium iodide solution when a further sample is shaken with the same solution of iodine in trichloromethane?

Directions for Questions 4 to 8 (See Rubric B)

4. When a few small crystals of iodine are placed in the bottom of a U-tube and chlorine passed over it, the iodine turns into a brown liquid with a brown vapour above it. On passing more chlorine through the tube, yellow crystals appear on the walls of the tube.

   If the chlorine supply is detached and the U-tube turned upside down and then upright again, the yellow crystals disappear and a brown vapour is seen. The reason for this is that
   A. the yellow crystals have reacted with the air
   B. the crystals have disintegrated
   C. the yellow crystals are very unstable. Inverting the tube supplies enough energy to decompose them
   D. the yellow crystals are present in an equilibrium mixture with chlorine and the brown vapour. Inverting the tube causes chlorine to be lost from the tube displacing the position of equilibrium and forming chlorine and brown vapour
   E. the yellow crystals are a compound of iodine and chlorine

5. On mixing solutions of calcium chloride and oxalic acid ($H_2C_2O_4$), a precipitate of calcium oxalate is formed and an equilibrium set up which can be represented by:

   $$\underset{\text{solution}}{CaCl_2} + \underset{\text{solution}}{H_2C_2O_4} \rightleftharpoons \underset{\text{solid}}{CaC_2O_4} + \underset{\text{solution}}{2HCl}$$

   Which of the following would dissolve the precipitate of calcium oxalate?
   A. Adding more calcium chloride solution
   B. Adding more oxalic acid
   C. Adding concentrated hydrochloric acid
   D. Stirring the mixture
   E. The addition of a catalyst

6. The equilibrium between iron and steam is represented by:

   $$3Fe(s) + 4H_2O(g) \rightleftharpoons Fe_3O_4(s) + 4H_2(g)$$

   This method has been used to prepare hydrogen. Which of the following will give an improved percentage conversion of steam to hydrogen?
   A. Adding a catalyst
   B. Blowing the steam more vigorously over the iron
   C. Carrying out the reaction at a higher pressure
   D. Carrying out the reaction in a closed vessel
   E. Using iron powder rather than iron filings

Equilibria

7. The equilibrium between silver nitrate and iron(II) sulphate is represented by:

$$Ag^+(aq) + Fe^{2+}(aq) \rightleftharpoons Ag(s) + Fe^{3+}(aq)$$

Which of the following would *not* shift the position of equilibrium towards the right?
A. Adding more silver nitrate solution
B. Adding a solution of another compound containing $Ag^+(aq)$ ions
C. Adding hydrochloric acid with which silver ions form a white precipitate
D. Dissolving the silver formed in nitric acid
E. Filtering off the deposit of silver

8. A beaker contains some water. Under the conditions existing at the time, the vapour pressure of the water is equal to the saturated vapour pressure of the air. Which of the following is *not* true?
A. No water molecules are escaping from the beaker because the air already holds as much water as it can at that temperature. A state of static equilibrium thus exists
B. Pure water always has a definite vapour pressure at a certain temperature providing that the atmospheric pressure is normal. This is because an equilibrium always exists between liquid water and water vapour
C. Cooling the beaker will cause condensation to appear on the sides of the beaker
D. As the temperature of the room rises the air can hold more water so the water in the beaker will start to evaporate
E. Some salt is dissolved in the water. The addition of ions from the salt makes the escape of water molecules from near the surface of the liquid more difficult so that the boiling point of the solution would rise above 100 °C

Directions for Questions 9 to 11 (See Rubric C)

9. When a reaction is at equilibrium,
  (i) the concentrations of the various substances remain unchanged provided that the external conditions remain constant
  (ii) the reactions between the various substances cease when the equilibrium point is reached
  (iii) changing the external conditions causes the equilibrium to be disturbed in such a way that the reaction itself tries to oppose the changes being made
  (iv) the amounts of each substance present are identical

10. The equilibrium reaction between bismuth chloride and water is represented by:
$$BiCl_3(aq) + H_2O(l) \rightleftharpoons BiOCl(s) + 2H^+(aq) + 2Cl^-(aq)$$
Correct deductions from this equation include
   (i) Bismuth chloride dissolved in concentrated hydrochloric acid gives a precipitate when a few cubic centimetres of water are added to the mixture
   (ii) Weak acids contain fewer $H^+(aq)$ ions than strong acids. They are therefore less effective in dissolving bismuth oxychloride than strong acids
   (iii) Bismuth chloride and bismuth oxychloride both dissolve in concentrated hydrochloric acid
   (iv) The addition of sodium chloride to a mixture of bismuth oxychloride and water has no effect on the position of equilibrium

11. When sulphur trioxide is heated in a closed tube it begins to decompose and eventually an equilibrium state is reached with sulphur trioxide, sulphur dioxide and oxygen all present in the tube together. When this state is reached
   (i) no more sulphur trioxide is decomposing
   (ii) no more sulphur dioxide and oxygen are combining together
   (iii) the mass of sulphur trioxide is equal to the combined masses of sulphur dioxide and oxygen
   (iv) the mass of sulphur trioxide present in the mixture remains constant

Directions for Questions 12 to 15 (See Rubric D)

Assertion		Reason
12. The yellow colour of bromine water is removed by adding alkali	BECAUSE	bromine and water form an equilibrium represented by: $Br_2(aq) + H_2O(l) \rightleftharpoons H^+(aq) + Br^-(aq) + HOBr(aq)$
13. Potassium chromate is yellow in alkaline solution and orange in acid solutions; the reaction is reversible and an equilibrium point can be found	BECAUSE	the equilibrium point is the half-way stage in a reaction

Equilibria

Assertion		Reason
14. When a solid is in equilibrium with its melt there is no change in the amounts of solid and liquid present	BECAUSE	when a solid is in equilibrium with its melt there is no transfer of particles from one phase to the other
15. When a precipitate of lead chloride containing radioactive lead is added to a saturated solution of lead chloride the radioactivity remains associated with the solid	BECAUSE	when solid lead chloride is added to a saturated solution of lead chloride there is no movement of particles from the solid into the liquid as the solution cannot hold any more solid

Chapter 20

ACIDS

Directions for Questions 1 to 10 (See Rubric A)

Questions 1 to 5

Five possible arrangements of molecules and ions are given below:
A. Widely separated molecules
B. Closely packed molecules with a random arrangement
C. Closely packed molecules with a regular arrangement
D. Widely separated ions
E. Closely packed ions

Which one of the above is in each case most likely to describe the arrangement of the particles in hydrogen iodide (melting point −50.8 °C; boiling point −35.4 °C)

1. at −100 °C

2. at −40 °C

3. at room temperature

4. when in solution in toluene

5. when in dilute solution in water

Questions 6 to 10

The graph in Figure 20.1 shows the change of pH as 0.1 M sodium hydroxide solution is added to 20 cm³ of 0.1 M acetic acid solution. Five points are marked on the curve.

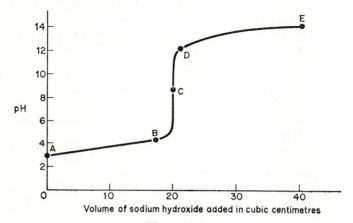

Figure 20.1

6. Which point gives the pH of 0.1 M acetic acid solution?

7. At which point is the pH changing most rapidly?

8. Which point is closest to the pH of pure water?

9. Which point represents the smallest concentration of hydrogen ions H^+ (aq)?

10. Which point gives the pH of the solution closest to the value at which a suitable indicator for following this reaction will change colour?

Directions for Questions 11 to 18 (See Rubric B)

11. Amongst a student's notes describing the tests that he carried out on a solution of hydrogen chloride in toluene, are the statements shown below. Pick out the statement that appears to be wrong.
    A. When two pieces of litmus paper, one piece taken from a desiccator and the other from a bottle that had been standing in the laboratory open to the air for some time, were put into the solution, the former was unaffected but the latter changed colour significantly
    B. Whereas a solution of hydrogen chloride in water reacted rapidly with marble chips, the solution in toluene had no effect on the substance
    C. As the solution does not conduct electricity there can be few, if any, ions present
    D. When the solution of hydrogen chloride in toluene was shaken with water in a separating funnel, two liquid layers became visible soon after the shaking was discontinued
    E. When the solution of hydrogen chloride in toluene is shaken with water (itself a poor conductor), the whole of the mixture remains non-conducting

12. Xylene is a liquid similar in properties to toluene. It does not mix with water. When a solution of hydrobromic acid (HBr) is shaken up with xylene and a portion of the xylene layer then tested, it is found to
    A. react with magnesium
    B. liberate carbon dioxide from carbonates
    C. give an acid reaction on testing with aqueous universal indicator solution
    D. contain no hydrogen bromide
    E. undergo electrolysis

13. When hydrogen chloride gas is dissolved in water, an equilibrium is set up represented by:

    $$HCl(g) + H_2O(l) \rightleftharpoons H_3O^+(aq) + Cl^-(aq)$$
    $$\text{I} \qquad \text{II} \qquad \text{III} \qquad \text{IV}$$

    Using the Roman numerals shown below each formula, which of these combinations represents two bases?
    A. I and III
    B. I and IV
    C. II and III
    D. II and IV
    E. III and IV

Acids

14. A weak acid is best described as
  A. a dilute acid
  B. an acid that is harmless
  C. an acid that does not easily donate protons
  D. an acid that reacts with very few other substances
  E. an acid present in fruit juice

15. Which of the following facts is true of sulphuric acid?
  A. It gives rise to salts called sulphides
  B. It is a reducing agent
  C. In the concentrated form it is a useful substance for drying gases
  D. The negative ions derived from sulphuric acid bear three charges
  E. It is manufactured by burning sulphur in air to form sulphur dioxide, followed by dissolving this gas in water

16. It is found that when a solution of barium hydroxide $(Ba(OH)_2)$ is added to 25 cm³ of a solution of nitric acid (HNO_3), neutralization occurs when 50 cm³ of barium hydroxide is added.
   It follows that
  A. the nitric acid is four times as concentrated as the barium hydroxide solution
  B. the nitric acid is twice as concentrated as the barium hydroxide solution
  C. the nitric acid has the same concentration as the barium hydroxide solution
  D. the barium hydroxide is twice as concentrated as the nitric acid solution
  E. the barium hydroxide is four times as concentrated as the nitric acid solution

17. A salt is defined as a substance produced by neutralizing an acid. Consequently
  A. salts always give a neutral solution on dissolving in water
  B. water is produced at the same time as the salt if the salt is being prepared by adding an acid to an alkali
  C. the method of preparing a salt from an acid and alkali is not possible if the acid is strong and the alkali weak
  D. neutralization is essentially a reaction between the ions causing acidity and the ions causing alkalinity, giving rise to a molecular compound. Salts, therefore do not contain ions
  E. a salt can only be made by a reaction involving an acid

18. Some substances are made by a method termed 'ionic association'. This method was used to prepare a sample of lead chloride. A solution of lead acetate, $(CH_3COO)_2Pb$, was added to 10 cm³ of 2 M potassium chloride solution. The lead chloride was precipitated as a white solid.

    The equation for the reaction is

    $$Pb^{2+}(aq) + 2Cl^-(aq) \rightarrow PbCl_2(s)$$

    What is the minimum quantity of 1 M lead acetate solution that must be used in this experiment to give the maximum yield of lead chloride?

    A. 2.5 cm³
    B. 5 cm³
    C. 10 cm³
    D. 20 cm³
    E. 40 cm³

Directions for Questions 19 to 23 (See Rubric C)

19. Universal indicator paper turns a colour indicating acidity when
    (i) it is laid on tartaric acid crystals
    (ii) it is placed in vinegar
    (iii) it is placed in dry glacial acetic acid
    (iv) it is placed in a solution of ammonium chloride

20. A solution may be assumed to be acidic if it
    (i) turns universal indicator paper red
    (ii) effervesces with magnesium ribbon
    (iii) conducts electricity, evolving hydrogen at the cathode
    (iv) dissolves sodium carbonate (washing soda)

21. Acidity
    (i) is measured by the pH scale
    (ii) is possessed only by substances which contain hydrogen
    (iii) is, in aqueous solution, a property of a chemical equilibrium in which hydronium ions are present
    (iv) can often be recognized by the occurrence of certain chemical reactions, for example, the liberation of carbon dioxide from carbonates

22. A certain solution of aluminium chloride in water reacts according to the equation

    $$Al(H_2O)_6{}^{3+}(aq) + H_2O(l) \rightleftharpoons [Al(H_2O)_5OH]^{2+}(aq) + H_3O^+(aq)$$

    The solution therefore
    (i) contains a hydrated cation
    (ii) is neutral to universal indicator
    (iii) contains two acids and two bases
    (iv) contains fewer H_3O^+ ions when more water is added

Acids

23. A chemist is studying the reaction between a base (formula XOH) and hydrochloric acid (HCl). Using 20 cm³ of a solution of the base, he finds that when 19 cm³ of the acid have been added to it, the pH is 8.0, but that when 21 cm³ of acid have been added, the pH is 2.5. Methyl orange changes colour between pH 2.9 and 4.6 and phenolphthalein between 8.3 and 10.0. Correct conclusions from this information are:
   (i) The base is weak but hydrochloric acid is a strong acid
   (ii) The concentrations of the base and the hydrochloric acid are the same
   (iii) Methyl orange would be preferable to phenolphthalein as an indicator for studying this reaction
   (iv) It is the X^+ ions present in the base that give it its basic properties

Directions for Questions 24 to 30 (See Rubric D)

Assertion		Reason
24. When hydrogen chloride gas is dissolved in toluene there is a considerable rise in temperature	BECAUSE	energy is required to break the bond between hydrogen and chlorine in hydrogen chloride
25. Hydrochloric acid is a good conductor of electricity	BECAUSE	solutions of all acids are fully ionized
26. Nitric acid and hydrochloric acid are called strong acids	BECAUSE	nitric acid and hydrochloric acid can be prepared in concentrated form
27. A strong acid makes universal indicator turn red	BECAUSE	a strong acid has a pH greater than 10
28. Water can behave both as an acid and as a base	BECAUSE	water is able to supply protons and accept protons
29. When a solution of sulphuric acid is added to a solution of barium hydroxide, the electrical conductivity reaches a minimum value at the end point	BECAUSE	when a solution of sulphuric acid is added to a solution of barium hydroxide, the pH of the mixture is a minimum at the end point

Assertion		Reason
30. Acids and bases react in the ratio 1 mole acid to 1 mole base	BECAUSE	H_3O^+ and OH^- ions react in the ratio 1 ion to 1 ion

Chapter 21

MOLECULES—BIG AND SMALL

Directions for Questions 1 to 6 (See Rubric A)

Questions 1 to 6
- A. Cracking
- B. Dehydrogenation
- C. Fermentation
- D. Hydrolysis
- E. Polymerization

Select the term appropriate to each of the following:

1. The treatment of starch with saliva

2. The essential process occurring in the production of beer

3. Passing the vapour of medicinal paraffin over strongly heated porous pot

4. The conversion of a saturated hydrocarbon molecule into an unsaturated hydrocarbon molecule with the same number of carbon atoms

5. The addition of an aqueous solution of 1,6-diaminohexane to a solution of adipyl chloride in carbon tetrachloride

6. The treatment of castor oil with sodium hydroxide to form a soap

Directions for Questions 7 to 11 (See Rubric B)

7. Carbohydrates are
   A. compounds composed of carbon atoms and water molecules arranged alternately
   B. compounds containing carbon and hydrogen
   C. compounds containing carbon, hydrogen, and oxygen, the hydrogen and oxygen being in the proportion 2 : 1
   D. compounds containing carbon and oxygen
   E. compounds containing starch.

8. When starch is broken down by the action of dilute acid which of the following statements does *not* apply to the result which occurs?
   A. Smaller molecules are produced
   B. If hydrolysis is complete the product will turn a solution of iodine dissolved in potassium iodide a deep blue colour
   C. The products will give a positive reaction to Fehling's test
   D. The product should be concentrated under reduced pressure before identification is carried out
   E. The final product glucose is, like starch, a carbohydrate

9. When glucose is converted into ethanol the best way to obtain a sample of fairly pure ethanol is by
   A. decanting the liquid away from the residue
   B. evaporating the mixture
   C. filtering the mixture
   D. distilling the mixture using a water condenser
   E. fractionally distilling the mixture using a fractionating column and a water condenser

10. Ethene (ethylene) is
    A. produced from ethanol by burning it in air
    B. industrially, still produced mainly from ethanol
    C. converted to ethanol by reaction with steam and a catalyst under pressure
    D. the main constituent of natural gas
    E. a comparatively unreactive compound and so has few uses

11. On heating fat in a frying pan until it smokes it is found that the fumes do not condense back to liquid fat on cooling to room temperature but remain as an unsaturated gas. The reason for this is that
    A. the fumes have polymerized.
    B. a new compound of higher molecular weight has been formed
    C. the fat has reacted with the air to form an unsaturated gas
    D. the fat has undergone hydrolysis
    E. the metal of the frying pan has acted as a catalyst for the 'cracking' of the fat molecules

Molecules—Big and Small

Directions for Questions 12 to 15 (See Rubric C)

12. In identifying the products of the breakdown of starch, paper chromatography may be used. In this method
    (i) it is necessary to use known substances as well for the purpose of comparison
    (ii) locating agents have to be applied to the chromatogram in order to make identification possible
    (iii) the samples being investigated are placed initially along a reference line drawn near the base of the paper and at right angles to the solvent flow
    (iv) the various substances present are carried along the paper to different points by the action of the solvent

13. When ethanol vapour is passed over broken porous pot a gas is produced which
    (i) burns with a smoky flame
    (ii) decolorizes bromine water
    (iii) decolorizes acidified potassium permanganate
    (iv) is an example of a saturated compound

14. Polythene is a plastic material made by polymerizing the hydrocarbon ethene (ethylene). Polythene is
    (i) solidified ethene
    (ii) composed of carbon and hydrogen only.
    (iii) an example of a monomer
    (iv) a compound, each molecule of which consists of a long chain-like structure

15. When depolymerizing Perspex it is necessary to
    (i) avoid inhaling the fumes evolved
    (ii) heat the test-tube containing the Perspex as strongly as possible
    (iii) condense the monomer produced
    (iv) add lauroyl peroxide as a catalyst

Directions for Questions 16 to 20 (See Rubric D)

Assertion		Reason
16. The degradation of starch by both dilute acid and enzyme action (using saliva) yields the same sugar as the final product	BECAUSE	on chewing a piece of bread for a long time it begins to taste sweet, showing that the starch is being broken down into a sugar in the body

Assertion		Reason
17. Man makes use of the energy of the sun through eating plants which contain glucose	BECAUSE	glucose is produced in plants by the process known as photosynthesis
18. Cracking, especially with the use of a catalyst, is an extremely important process in the oil industry	BECAUSE	crude oil is largely composed of heavy hydrocarbon molecules which are less useful than the lighter molecules obtained by cracking
19. A detergent can be made by treating fats or oils with concentrated sulphuric acid	BECAUSE	a detergent is obtained from a carbohydrate using concentrated sulphuric acid, the acid absorbing water from the molecule
20. When a polymer is formed from two different monomers it is necessary that the monomers have a reactive group of atoms at both ends of the molecules	BECAUSE	when a polymer is formed the monomers join together to form a long chain of molecules

Chapter 22

AMMONIA

Directions for Questions 1 to 3 (See Rubric A)

Questions 1 to 3
  A. Calcium oxide   D. Potassium sulphate
  B. Copper(II) oxide   E. Soda lime
  C. Hydrogen chloride

Choose from the above list the substance applicable to each of the following:

1. This substance is normally used for drying ammonia gas

2. This substance forms dense white fumes on coming into contact with ammonia gas

3. This substance supplies one of the three major plant foods

Directions for Questions 4 to 8 (See Rubric B)

4. Which of the following is *not* a protein?
   A. Cheese   D. Meat
   B. Hair   E. Sugar
   C. Horn

5. Ammonia contains the elements nitrogen and hydrogen into which it is easily decomposed. Which of the following pieces of information would be the most useful for determining the formula of ammonia (assuming that all volumes are measured under the same conditions)?
   A. The mass of nitrogen and hydrogen produced from a known volume of ammonia
   B. The mass of ammonia which is necessary to produce one mole of hydrogen atoms, H
   C. The mass of ammonia which is necessary to produce one mole of hydrogen molecules, H_2
   D. The volumes of nitrogen and hydrogen produced from a known mass of ammonia
   E. The volumes of nitrogen and hydrogen produced from a known volume of ammonia

6. Which of the following is *not* a correct statement concerning ammonia?
   A. Ammonia is not very soluble in water
   B. Ammonia turns moist red litmus paper blue
   C. Ammonia is less dense than air
   D. Ammonia can be produced by warming an ammonium compound with an alkali
   E. Ammonia is used, in the anhydrous state, for direct injection of nitrogen into the soil

7. The nitrogen cycle is shown in outline in Figure 22.1

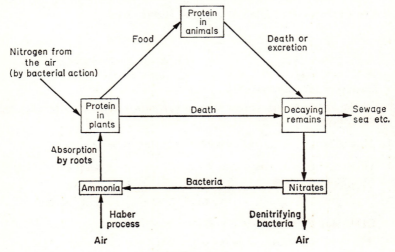

Figure 22.1

Which of the following is *not* a correct deduction?
   A. The order of 'nitrates' and 'ammonia' should be reversed
   B. In nature the small quantity of nitrogen lost to the air is balanced by the conversion of atmospheric nitrogen into plant protein by bacterial means
   C. Where land is cultivated the loss of nitrogen from the soil far exceeds the return
   D. Animals cannot make use of nitrogen direct from the air
   E. Nitrogen is the only essential plant food

Ammonia

8. Assuming that the only consideration is the maximum amount of nitrogen to be applied at the lowest cost, which of these fertilizers would you recommend to a farmer as the best buy?

Fertilizer	Chemical name	Formula	Mass of 1 mole (g)	Price per kilogramme
A. Nitrate of potash	Potassium nitrate	KNO_3	101	12p
B. Nitrate of soda	Sodium nitrate	$NaNO_3$	69	9p
C. Nitro-chalk	⅔ ammonium nitrate	NH_4NO_3	80	8p
	⅓ calcium carbonate	$CaCO_3$	100	
D. Sulphate of ammonia	Ammonium sulphate	$(NH_4)_2SO_4$	132	7p
E. Urea	Urea	$CO(NH_2)_2$	60	50p

Directions for Questions 9 to 12 (See Rubric C)

9. Which of the following observations would be made if the experiment illustrated in Figure 22.2 were carried out?

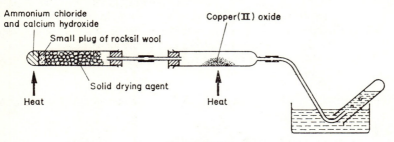

Figure 22.2

(i) An alkaline gas is collected
(ii) The gas collected does not burn
(iii) The copper(II) oxide remains unchanged (it is found to be a catalyst to the reaction occurring)
(iv) Drops of water form on the cool parts of the combustion tube

10. Dry ammonia is passed over heated iron wool and the products obtained are then passed over heated copper(II) oxide. It is found that
    (i) after passing over the iron wool the volume of gases present in the mixture on cooling is found to be the same as the initial volume of gas
    (ii) after passing over the iron wool a mixture of two gases is produced containing equal amounts of each
    (iii) when the final gaseous product is cooled to room temperature the volume of nitrogen is one quarter that of the volume of ammonia gas used in the reaction
    (iv) when the final gaseous product is cooled to room temperature its volume is half that of the volume of ammonia gas used in the reaction

    (All measurements are made under the same conditions of temperature and pressure)

11. The Haber process for the production of ammonia is one of the most important of all industrial processes. Among its features are
    (i) the use of an iron catalyst
    (ii) the essential reaction is a reversible reaction so conditions have to be chosen carefully to give the best yield of ammonia
    (iii) the conversion of nitrogen from the air into a substance from which nitrogenous fertilizers can easily be produced
    (iv) a good yield of ammonia is obtainable even without the use of high pressure

12. A widely used fertilizer is sulphate of ammonia (ammonium sulphate, $(NH_4)_2SO_4$.
    Among its properties are:
    (i) It is easily produced by neutralizing sulphuric acid with ammonia solution
    (ii) It is readily soluble
    (iii) It contains approximately 20% of nitrogen
    (iv) It has no effect on the pH value of soil

Directions for Questions 13 to 15 (See Rubric D)

Assertion		Reason
13. When a protein is heated with soda lime, the gases produced react alkaline to universal indicator but, after passing through 'activated carbon', the gases are no longer alkaline	BECAUSE	'activated carbon' is able to absorb alkaline gases only

Assertion		Reason
14. The ammonium ion is acidic	BECAUSE	in the equilibrium $NH_3(aq) + H_3O^+(aq) \rightleftharpoons NH_4^+(aq) + H_2O(l)$ the equilibrium position is normally well to the right since ammonia is a stronger base than water
15. An important use of ammonia is in the production of explosives	BECAUSE	one of the constituent elements of ammonia, hydrogen, forms explosive mixtures with air or oxygen

Chapter 23

ENERGY

Directions for Questions 1 to 5 (See Rubric A)

Questions 1 to 5

    A. The heat of combustion   D. The energy of a chemical
    B. The heat of reaction   E. The 'work' of a reaction
    C. The heat of vaporization

From the above list choose the correct term for each of the following:

1. This is the heat change occurring when 1 mole of a substance is completely burnt in oxygen

2. The absolute value of this quantity cannot be measured

3. This expression is given the symbol ΔG

4. The algebraic sign of this term for a particular reaction indicates whether the chemical change is in fact possible

5. This term is also equal to the limiting energy of a reaction

Directions for Questions 6 to 11 (See Rubric B)

6. The heat of combustion of carbon is determined by finding the rise in temperature produced in a known quantity of water when a definite mass of carbon is completely burnt in oxygen. The results obtained in such an experiment were:

mass of dry charcoal before experiment = 0.595 g
mass of ash remaining after experiment = 0.015 g
mass of water used = 580 cm³

1 J of energy raises the temperature of 1 g of water by 0.24 °C
The heat of combustion of 1 mole of carbon atoms is therefore:

A. $\dfrac{7.5 \times 580}{0.58 \times 0.24}$ J mol^{-1}

B. $\dfrac{7.5 \times 580 \times 12}{0.58 \times 0.24}$ J mol^{-1}

C. $\dfrac{7.5 \times 580 \times 0.58}{12 \times 0.24}$ J mol^{-1}

D. $\dfrac{7.5 \times 580}{12 \times 0.58 \times 0.24}$ J mol^{-1}

E. $\dfrac{580 \times 12}{7.5 \times 0.58 \times 0.24}$ J mol^{-1}

7. A certain car does 14 kilometres per cubic decimetre of petrol. Petrol may be regarded as being mainly pentane, C_5H_{12}. In a cubic decimetre of petrol there are approximately 620 g of pentane and, when 1 mole is completely burnt, 3.24 kJ of heat are evolved. The energy liberated when the car travels at a distance of 7 km is therefore:

A. $\dfrac{620 \times 3.24}{2}$ kJ

B. 620×3.24 kJ

C. $\dfrac{620 \times 3.24}{72 \times 2}$ kJ

D. $\dfrac{620 \times 3.24}{72}$ kJ

E. $\dfrac{72 \times 3.24}{620 \times 2}$ kJ

8. The heat of reaction for the reaction between potassium chloride solution and silver nitrate solution may be determined by mixing the solutions in a polythene bottle fitted with stopper and thermometer.

   Which of the following does *not* apply to this experiment?
   A. The final volume of solution must be 1 dm^3
   B. The solutions and the apparatus are left on the bench for some time before use.
   C. The solutions must be shaken well before recording the highest temperature reached on the thermometer
   D. The amount of heat used in warming up the polythene container is negligible
   E. There is negligible heat loss from a polythene bottle

9. A solution of 500 cm^3 of 2 M potassium hydroxide is added to 500 cm^3 of 2 M hydrochloric acid and the mixture well stirred. The rise in temperature (T_1) is recorded. The experiment is then repeated using 250 cm^3 of each solution and the rise in temperature (T_2) is recorded. It is found that
   A. T_1 is four times as large as T_2
   B. T_1 is twice as large as T_2
   C. T_1 is equal to T_2
   D. T_2 is twice as large as T_1
   E. T_2 is four times as large as T_1

10. The equation for the reaction between hydrogen and bromine is
    $$H_2(g) + Br_2(g) \rightarrow 2HBr(g)$$
    The energy required to break one mole of hydrogen molecules, H_2, into atoms is 435 kJ and the energy required to break one mole of bromine molecules, Br_2, into atoms is 224 kJ. The energy required to break one mole of hydrogen bromide molecules, HBr, into separate atoms is 366 kJ.
    The energy change for this reaction is therefore
    A. $+1391$ kJ   D. -73 kJ
    B. $+293$ kJ   E. -1391 kJ
    C. $+73$ kJ

11. A very simple fuel cell can be made by electrolysing a 1 M solution of sodium hydroxide in the apparatus shown in Figure 23.1. On disconnecting the d.c. supply and putting a high-resistance

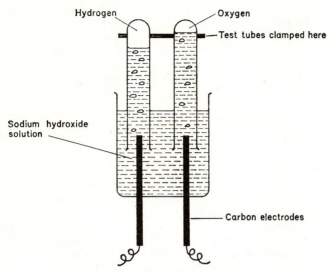

Figure 23.1

voltmeter across the carbon electrodes a voltage is recorded. Which of the following statements is *incorrect*?
A. A d.c. supply is not necessary; a low-voltage a.c. supply could also be used
B. The reason for electrolysing an alkaline solution to produce hydrogen and oxygen is that some carbon dioxide is produced by oxidation of the electrodes and this is absorbed by the alkali
C. The gases absorbed on the carbon electrodes provide the reactants for the cell reaction resulting in the production of a voltage
D. A fuel cell is a method of converting energy from oxidation of a fuel directly into electricity; in this simple cell the essential reaction is

$$2H_2(g) + O_2(g) \rightarrow 2H_2O(l)$$

E. A fuel cell has a higher efficiency for utilizing the energy transformed during the reaction than have conventional sources of energy

Multiple-Choice Questions in 'O' Level Chemistry

Directions for Questions 12 to 15 (See Rubric C)

12. A known mass of water is placed in a metal can and heated by an improvised spirit lamp which is lit after asbestos squares have been placed around the apparatus as a draught shield. The spirit lamp contains propan-1-ol and the rise in temperature of the water is measured whilst a known mass of alcohol is burnt. The experiment is carried out in order to determine the heat of combustion of the alcohol. The value obtained is considerably less than that quoted in a book of data. Which of the following factors contribute towards the low value obtained?
    (i) Some heat is absorbed by the vessel containing the water
    (ii) Some soot is deposited on the wick of the lamp
    (iii) Incomplete burning of some of the alcohol leads to the formation of carbon monoxide
    (iv) The asbestos squares do not cut out all the draughts

13. In determining the energy change in making an aqueous solution, use is made of the facts that
    (i) for solutions of not more than 1.0 M concentration it requires approximately 4.18 kJ to warm 1 dm^3 of solution by 1 °C.
    (ii) for solutions of not more than 1.0 M concentration the temperature change in degrees Celsius is numerically equal to the energy change in kilojoules divided by 4.18
    (iii) if a thermometer is held in the centre of the solution, an accurate record of the temperature change occurring is obtained because the thermometer is surrounded by solution of very nearly the same temperature and so heat losses are negligible
    (iv) if the water cools down during the dissolving process the energy change for this process is given a negative sign

14. Consider the energy change occurring when ammonium nitrate is dissolved in water:

    $NH_4NO_3(s) + aq \rightarrow 1.0$ M $NH_4NO_3(aq)$; $\Delta H = +25.1$ kJ

    It follows that
    (i) the reaction gives out heat
    (ii) the aqueous solution of ammonium nitrate has more energy than the separate substances
    (iii) making 2 dm^3 of a 1 M solution of ammonium nitrate will produce a total energy change of $\Delta H = +50.2$ kJ
    (iv) the energy level diagram for the reaction is

    $$\begin{array}{l} \quad\quad\quad\quad\quad\quad\quad \text{- - - - 1.0 M } NH_4NO_3(aq) \\ \quad\quad\quad\quad\quad\quad\quad \uparrow \\ NH_4NO_3(s) + aq \text{ - - - - } \Delta H = +25.1 \text{ kJ} \end{array}$$

15. The essential reaction of the Daniell Cell is

$$Zn(c) + Cu^{2+}(aq) \rightarrow Zn^{2+}(aq) + Cu(c) \qquad \Delta G = -211.7 \text{ kJ}$$
$$\Delta H = -217.8 \text{ kJ}$$

The energy change, ΔG, is the limiting electrical energy for this reaction. ΔH is the heat of reaction. This information implies that
   (i) there is a definite maximum e.m.f. available from such a cell
   (ii) not all the thermal energy available from the chemical reaction can be used as electrical energy
   (iii) some heat is evolved in the cell
   (iv) the principle of Conservation of Energy does not apply to chemical reactions

Directions for Questions 16 to 20 (See Rubric D)

Assertion		Reason
16. As the number of carbon atoms present in an alcohol increases, the more easily does it burn	BECAUSE	the heat of combustion of an alcohol becomes greater as the number of carbon atoms per molecule increases
17. In an exothermic reaction the heat of reaction ΔH is given a positive sign	BECAUSE	in an exothermic reaction heat is evolved
18. When 25 cm³ of 1 M silver nitrate solution and 25 cm³ of 1 M potassium chloride solution are mixed, ΔH for the precipitation of silver chloride is found to have the same value as that obtained when 25 cm³ of 1 M silver nitrate solution is added to 25 cm³ of 1 M sodium chloride solution	BECAUSE	sodium and potassium are in the same group of the periodic table so that any energy change associated with a sodium compound is almost identical to that associated with the corresponding potassium compound
19. Mechanical energy obtained from a steam engine and electrical energy obtained from a cell are both obtained by the transference of heat energy into another form of energy	BECAUSE	a steam engine and an electric cell both depend upon chemical reactions, and in the course of the reaction energy redistribution takes place

Assertion		Reason
20. The heat change for a reaction is essentially the difference between the energy that has to be supplied to a reaction to break chemical bonds and the energy lost to the surroundings through the making of new bonds	BECAUSE	the energy of a chemical is equal to the energy of bonding holding the atoms together in the particles of the compound

Chapter 24

RADIO-CHEMISTRY

Directions for Questions 1 to 3 (See Rubric A)

Questions 1 to 3
    A. α-rays   D. X-rays
    B. β-rays   E. cosmic rays
    C. γ-rays

1. Which of these rays is identical with electrons?

2. Which of these rays is emitted by an atom of $^{231}_{91}$Pa when it changes into an atom of $^{227}_{89}$Ac?

3. Which of these rays gives rise to tracks on a nuclear emulsion photographic plate taken up a mountain and subsequently developed?

Directions for Questions 4 to 7 (See Rubric B)

4. ^{212}Bi has a half-life of about 60 min. The time taken for the radioactivity to decay to one sixty-fourth of its initial value is
    A. just under one minute   D. 32 hours
    B. 6 hours   E. 64 hours
    C. 8 hours

5. The thorium decay series is

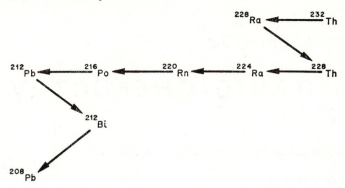

To determine the half-life of ^{212}Pb a few drops of lead nitrate solution and barium nitrate solution are added to a solution of thorium nitrate. The lead and thorium hydroxides are then precipitated by the addition of concentrated ammonia solution. The barium ions are added as a 'holdback carrier'. This means that

A. the barium ions are slowing down the decay of the lead isotope making it easier to investigate
B. the barium ions are preventing precipitation of other radioactive material such as ^{224}Ra
C. the barium ions are acting as a negative catalyst for the reaction between ammonia and the lead and thorium ions
D. the barium ions allow only radioactive matter to be precipitated
E. the barium ions reduce the solubility of the lead and thorium hydroxides

6. An isotope A decays to give a daughter B which is also radioactive. The half-life of isotope A is greater than that of isotope B. The graph of total activity (y axis) against time (x axis) has the shape shown in Figure 24.1.

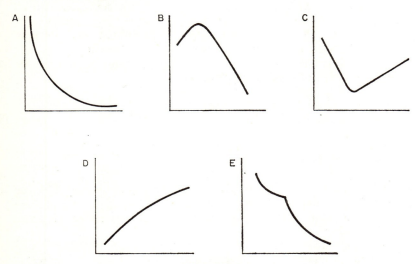

Figure 24.1

7. Which of the following is *not* a suitable use for a radioactive isotope?
   A. accurate dating of a seventeenth-century wood carving
   B. detection of leaks in a buried pipe
   C. the measurement of the solubility of an almost insoluble substance
   D. the measurement of the rate of uptake of trace elements by plants
   E. to show that chemical equilibrium is a dynamic state

Directions for Questions 8 to 11 (See Rubric C)

8. The quantity of radiation emitted by a radioactive isotope is
   (i) dependent on the temperature; the higher the temperature the more readily it will give out radiation
   (ii) dependent on the amount of carrier mixed with the isotope
   (iii) greater, the higher the energy of the particles being emitted
   (iv) proportional to the concentration of the isotope when in solution

9. Soddy reported that lead obtained from the mineral pitch-blende had an atomic mass of 206 but that lead obtained from the mineral thorianite had an atomic mass of 208.4. This is because
   (i) his figures were later shown to contain experimental errors
   (ii) the element lead is the one exception to the statement that all atoms of the same element are identical in every respect
   (iii) the sample of lead obtained from the thorianite was not pure and contained atoms of another, heavier element
   (iv) lead from thorianite contains a higher proportion of a heavier isotope than lead from pitch-blende

10. Isotopes of the same element have
    (i) different atomic numbers
    (ii) different mass numbers
    (iii) atoms with the same number of neutrons
    (iv) atoms with the same number of protons

11. If a nuclear photographic plate is left in a darkroom in a dilute solution of thorium nitrate for a few days, developed and then examined under a microscope, forked tracks are seen on the plate. The tracks are
    (i) due to radioactive decay of a thorium isotope ($^{228}_{90}Th$)
    (ii) three-dimensional, but occasionally a track is visible which is more or less in the plane of the emulsion of the plate
    (iii) able to provide information about the energy of the particles producing them
    (iv) forked, due to subsequent decay of the daughter formed from $^{228}_{90}Th$

Directions for Questions 12 to 15 (See Rubric D)

Assertion   **Reason**

12. A fast photographic plate, when placed in complete darkness beside crystals of a radioactive chemical, becomes slowly 'fogged' BECAUSE all radioactive chemicals are phosphorescent and continue to glow in the dark for a considerable time after removal from a light source

13. Rubber or polythene gloves should always be used for handling radioactive specimens BECAUSE gloves made from rubber or polythene protect the wearer from radiation

Radio-chemistry

	Assertion		Reason
14.	When equal volumes of 1 M solutions of potassium hydroxide and hydrochloric acid are counted separately and the solutions then mixed, a marked change in counting rate occurs on mixing	BECAUSE	when equal volumes of 1 M solutions of potassium hydroxide and hydrochloric acid are mixed, two new substances, potassium chloride and water are formed which emit a different total quantity of radiation
15.	In radiochemical experiments a carrier always has to be used	BECAUSE	in radiochemical experiments the radioactive isotope used is employed only in very small quantities